"There are only three kinds of disease: made wrong, bad luck, worn out!"

Dr Eugene Allison Stead

CARDIAC MRI IN THE DIAGNOSIS, CLINICAL MANAGEMENT AND PROGNOSIS OF ARRHYTHMOGENIC RIGHT VENTRICULAR CARDIOMYOPATHY/DYSPLASIA

ELSEVIER *science &*
technology books

Companion Web Site:
www.store.elsevier.com/9780128012833

Cardiac MRI in the Diagnosis, Clinical Management and Prognosis of Arrhythmogenic Right Ventricular Cardiomyopathy/Dysplasia
Aiden Abidov, Isabel B. Oliva, Frank I. Marcus, Editors

Available Resources:

Extra content and support files.

TOOLS FOR ALL YOUR TEACHING NEEDS
textbooks.elsevier.com

ACADEMIC
PRESS

CARDIAC MRI IN THE DIAGNOSIS, CLINICAL MANAGEMENT AND PROGNOSIS OF ARRHYTHMOGENIC RIGHT VENTRICULAR CARDIOMYOPATHY/ DYSPLASIA

Edited by

AIDEN ABIDOV

ISABEL B. OLIVA

FRANK I. MARCUS
*Department of Medicine/Division of Cardiology
and Department of Medical Imaging,
University of Arizona, Tucson, AZ, USA*

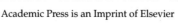

AMSTERDAM • BOSTON • HEIDELBERG • LONDON
NEW YORK • OXFORD • PARIS • SAN DIEGO
SAN FRANCISCO • SINGAPORE • SYDNEY • TOKYO

Academic Press is an Imprint of Elsevier

ELSEVIER

Academic Press is an imprint of Elsevier
125 London Wall, London EC2Y 5AS, UK
525 B Street, Suite 1800, San Diego, CA 92101-4495, USA
50 Hampshire Street, 5th Floor, Cambridge, MA 02139, USA
The Boulevard, Langford Lane, Kidlington, Oxford OX5 1GB, UK

British Library Cataloguing-in-Publication Data
A catalogue record for this book is available from the British Library

Library of Congress Cataloging-in-Publication Data
A catalog record for this book is available from the Library of Congress

ISBN: 978-0-12-801283-3

For information on all Academic Press publications
visit our website at http://store.elsevier.com/

Working together
to grow libraries in
developing countries

www.elsevier.com • www.bookaid.org

Publisher: Mica Haley
Acquisition Editor: Stacy Masucci
Editorial Project Manager: Sam W. Young
Production Project Manager: Chris Wortley
Designer: Matthew Limbert

Typeset by Thomson Digital

Printed and bound in United States of America

Dedication

To my dear wife, Yulia: thank you for always being there for me, being my pillar of strength, and supporting me through anything and everything I aspired to achieve.

To my dear kids Elnur, Amir, Meira, and Dan: thank you for always inspiring me and making me strive to be the best dad I could be. I love you all so much.

Aiden Abidov

To my caring husband, Felipe: You are the love of my life! Thank you for your continuous support and love, you make me a better person.

To my little Sophia: You are my life, we love you more than anything in this world.

To my parents, brother, and sister: Thank you for your love and for raising me to be the best I can be. Your successes have always inspired me; I miss you all every day!

Isabel Oliva

To my understanding wife, Janet who has tolerated her workaholic husband for many years.

Frank I. Marcus

Acknowledgment

The authors are indebted to Mrs Yvette M. Barnes, MEd for technical assistance in the preparation and submission of the manuscript.

Contents

6. Association of Phenotype and Genotype in the Diagnosis
and Prognosis of ARVC/D in the Adult Population

AMIT PATEL, LUISA MESTRONI, FRANK I. MARCUS

7. Diagnostic Evaluation of Children with Known
or Suspected ARVC/D

FRANK I. MARCUS, AIDEN ABIDOV

8. Differential Diagnosis of ARVC/D

AIDEN ABIDOV, FRANK I. MARCUS

9. Special Cases and Special Populations: Tips and Tricks
to Obtain a Diagnostic CMR

ISABEL B. OLIVA, AIDEN ABIDOV

10. Prognostic Value of Cardiac MRI in ARVC/D

ISABEL B. OLIVA, AIDEN ABIDOV

List of Contributors

Aiden Abidov Department of Medicine/Division of Cardiology and Department of Medical Imaging, University of Arizona, Tucson, AZ, USA

Cristina Basso Department of Cardiac, Thoracic and Vascular Sciences, University of Padua Medical School, Padua, Italy

Maarten J. Cramer Department of Cardiology, University Medical Center Utrecht, Utrecht, The Netherlands

Pieter A. Doevendans Department of Cardiology, University Medical Center Utrecht, Utrecht, The Netherlands

Arun Kannan Department of Medicine/Division of Cardiology and Department of Medical Imaging, University of Arizona, Tucson, AZ, USA

Frank I. Marcus Department of Medicine/Division of Cardiology and Department of Medical Imaging, University of Arizona, Tucson, AZ, USA

Thomas P. Mast Department of Cardiology, University Medical Center Utrecht, Utrecht, The Netherlands

Luisa Mestroni Cardiovascular Institute, University of Colorado Anschutz Medical Campus, Aurora, CO, USA

Isabel B. Oliva Department of Medicine/Division of Cardiology and Department of Medical Imaging, University of Arizona, Tucson, AZ, USA

Ahmed K. Pasha Department of Medicine/Division of Cardiology and Department of Medical Imaging, University of Arizona, Tucson, AZ, USA

Amit Patel Cardiovascular Institute, University of Colorado Anschutz Medical Campus, Aurora, CO, USA

Kalliopi Pilichou Department of Cardiac, Thoracic and Vascular Sciences, University of Padua Medical School, Padua, Italy

Stefania Rizzo Department of Cardiac, Thoracic and Vascular Sciences, University of Padua Medical School, Padua, Italy

Arco J. Teske Department of Cardiology, University Medical Center Utrecht, Utrecht, The Netherlands

Gaetano Thiene Department of Cardiac, Thoracic and Vascular Sciences, University of Padua Medical School, Padua, Italy

1

Introduction

Frank I. Marcus, Aiden Abidov

Department of Medicine/Division of Cardiology
and Department of Medical Imaging,
University of Arizona, Tucson, AZ, USA

This book aims to evaluate the role of the MRI in the diagnosis, clinical management, and prognosis of arrhythmogenic right ventricular cardiomyopathy/dysplasia (ARVC/D). You may ask "Isn't this too narrow a focus for this rare disease?" Let us evaluate this concern.

First, ARVC/D is now more frequently diagnosed as it is becoming better known. It is estimated that it occurs in 1:5000 individuals but it may be present in a higher incidence since one may have a pathological gene for this disease yet have little or no clinical manifestations. This is known as a lack of association of genotype and phenotype. Thus, ARVC/D may be a less rare disease than is presently thought. In addition, since it is a cause of sudden cardiac death, particularly in the young, it is important to be able to recognize it in order to prevent this catastrophic event.

Another question is, why should we focus our attention on one imaging modality, the MRI, particularly when this imaging modality is more expensive and less readily available than 2D echocardiography? In contrast to echocardiography, an MRI can provide more accurate quantitative evaluation of the right ventricular function and structure. Specifically, it can accurately access right ventricular ejection fraction as well as segmental wall motion abnormalities of the right ventricle. Based on hundreds of published papers, cardiac MRI is a useful diagnostic imaging modality in patients suspected of having ARVC/D and is particularly valuable since important limitations of MRI (such as the need for breath-holding, inability to scan patients with permanent pacemakers or ICDs, etc.) have largely been overcome. The finding of abnormal right ventricular function or structure by 2D echocardiogram in a patient suspected of having ARVC/D should be confirmed by MRI since the latter is more reliable for the diagnosis. It is also important that the radiologist/cardiologist who is interpreting the MRI should be aware of normal variants of the right ventricular contractility patterns, particularly that of an apparent bulging of the right ventricular free wall at the insertion of the right ventricular papillary muscle.

Other questions include the age at which ARVC/D is manifest. Should an MRI be done in children who have the genetic abnormality but no clinical manifestation of the disease? How rapidly do the abnormalities of the RV change in this disease? This would determine how frequently the MRI should be reassessed in first-degree relatives who may have no or minimal symptoms.

An important consideration is the increased safety of MRI, especially absence of exposure to ionizing radiation and nephrotoxic iodine contrast. This allows sequential MRI studies in young patients without increased associated risk of imaging. Excellent spatial resolution and safety of cardiac MRI makes it an ideal methodology for follow-up of patients with known or suspected ARVC/D.

Finally, the MRI is useful in differential diagnosis that includes several conditions mimicking ARVC/D, such as cardiac sarcoidosis, left-to-right supraventricular shunts, and myocarditis. Also, in some cases, myocardial–pericardial adhesions can cause abnormal right ventricular wall motion. The use of gadolinium contrast to detect and localize scar/fibrosis in the left or right ventricular myocardium is unique to MRI, as is the ability of cardiac MRI to provide effective tissue characterization, including fibro-fatty infiltration, inflammation, thrombosis, etc.

Recent developments in the field of advanced echocardiography, cardiac CTA, and nuclear cardiology have many interesting applications that could significantly enhance the armamentarium of physicians in the diagnosis and management of ARVC/D. In this book, we included a brief overview of novel non-MRI-based imaging methodologies that are useful in this disease.

In summary, there are many important clinical areas of interest reflecting the role of the MRI and other rapidly developing cardiac imaging methodologies in patients with ARVC/D. In our book, we provide our readers with a convenient overview of these areas.

However, there are three types of problems with cardiac imaging in general, and cardiac MRI in particular for the evaluation of ARVC/D. (Fig. 1.1):

1. **The problem of ordering the right test for the patient's age and clinical presentation.** Patients with known or suspected ARVC/D are a highly heterogeneous group and include patients with confirmed ARVC/D, asymptomatic gene carriers, and relatives of patients with ARVC/D, as well as patients with suspected or possible ARVC/D. There is significant disagreement about which test should be utilized in these populations, which one is the most effective for screening, and whether the layered testing concept should be considered in the

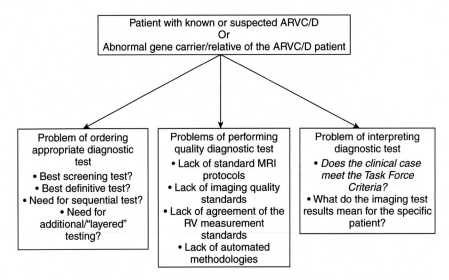

FIGURE 1.1 **Problems encountered in evaluation and management of patients with known or suspected ARVC/D.**

"borderline" cases. The Modified Task Force criteria focused on the specificity of echo and MRI measurements of ARVC/D and possibly at the expense of sensitivity, particularly of early or clinically "silent" disease cases.

2. **The problem of performing a good-quality diagnostic MRI.** For years, we have been reviewing MRI studies of patients with either known or probable disease performed in imaging laboratories from many centers in the United States and abroad. There is marked variability of the diagnostic quality of these studies. Also, there are many MRI protocols utilized in different centers. Current lack of standardization in MRI protocols for the ARVC/D patients is concerning. There is an urgent need to improve this situation.

3. **The problem of interpreting results of the MRI study.** Even negative results in particular clinical populations may mean just one negative diagnostic criterion among many others that must be considered in such a complex diagnosis as ARVC/D. At times the decision-making process is based completely on the imaging study. A false-negative study can be associated with increased risk, and a false-positive test may dramatically change the patient's life and have long-lasting consequences both for the patient and for the society. One of these situations we have encountered is implantation of ICDs in young patients who have borderline tests, or tests that are negative but are interpreted as positive even though other diagnostic tests were

not considered. Recently, with the development of genetic testing, interpretation of imaging test results in association with genetic defects in asymptomatic individuals raises important clinical decisions. These patients may be subjected to changes in their occupation, limitations in their athletic activities (such as college sports) and lifestyle even though their anatomic data do not suggest an increased risk.

In this book, we address these problems and provide quick access to evidence-based algorithms and methods utilized currently in the state-of-the-art imaging laboratories. We have utilized an exhaustive literature search, but we also give readers flow diagrams, clinical algorithm schemes, and figures. Easy access to these data may save time and effort in reaching important clinical decisions and utilize an important principle of the modern imaging: "The right test for the right patient."

This book provides a quick reference to assist with standardization of the imaging protocols, particularly for the practicing imagers and clinicians who may encounter patients with known or suspected ARVC/D. This book is designed to be user friendly. We provide clinical examples as well as online tools and videos to illustrate interesting cases from our practice.

We hope to have the readers' feedback and maintain online communication with interested clinicians and researchers in order to further enhance the potential of cardiac MRI and other imaging modalities in the diagnosis and management of ARVC/D.

Arrhythmogenic Cardiomyopathy: History and Pathology

Gaetano Thiene, Stefania Rizzo,
Kalliopi Pilichou, Cristina Basso

Department of Cardiac, Thoracic and Vascular Sciences,
University of Padua Medical School, Padua, Italy

INTRODUCTION

Arrhythmogenic right ventricular cardiomyopathy dysplasia (ARVC/D) is a life-threatening entity, which has drawn the attention of the scientific community for the last 30 years since it is a significant cause of premature death [1,2]. Young people, especially athletes, may die suddenly because of abrupt lethal cardiac arrhythmias, namely ventricular fibrillation, precipitated by exercise [3]. The present chapter will deal with some aspects of the disease: history, terminology, biological background, pathology, and morphological criteria for diagnosis, endomyocardial biopsy, and recapitulation of the disease in transgenic mice.

HISTORY

It is a "rediscovered" disease, since its knowledge dates back centuries. The early description belongs to the pathologist Giovanni Maria Lancisi, who first described its heredofamilial peculiarity [4]. In a chapter on hereditary predisposition to cardiac aneurysms and bulgings in his book *De Motu Cordis et Aneurysmatibus* (on the movements of the heart and aneurysms), he reported the history of a family with disease recurrence in four generations, featured by cardiac palpitations and sudden death.

FIGURE 2.1   **First historical description of ARVC/D in the book** *De Motu Cordis et Aneurysmatibus* **published in 1736 by Giovanni Maria Lancisi, Professor of Anatomy in Rome and Pope's Physician.** *Courtesy of Arnold Katz.*

Dilatation and aneurysms of the right ventricle (RV), which filled the right chest, were observed at autopsy (Fig. 2.1).

René Laennec, the French doctor and inventor of the stethoscope, in the book *"De l'auscultation mediate ou traite' du diagnostic des maladies des poumons et du Coeur"* (on the mediated auscultation and treatise of the diagnosis of lung and heart disease) published in 1819, first drew attention to the relationship between fatty tissue in the right ventricle (RV) and sudden death [5]. The walls were described as extremely thin "especially at the apex of the heart and the posterior side of the right ventricle." The risk of sudden death in a fatty heart was confirmed by the protagonist Dr. Lydate in Middlemarch of George Eliot in 1871, who, talking to his patient, said, "You are suffering from what is called fatty degeneration of the heart… it is my duty to tell you that death from the disease is often sudden…" [6]. In 1905 William Osler, in his famous treatise "The Principles and Practice of Medicine", described a case of a 40-year-old man, who died suddenly while climbing up a hill. The heart showed biventricular massive myocardial atrophy with very thin walls, as to be named "parchment heart" [7]. The specimen is now part of the Abbott collection in Montreal and was reviewed by Segall in 1950 [8].

In 1952, Uhl reported the fatal case of an infant, which has been the source of misconception and controversy [9]. A female infant died at the age of 8 months, with congestive heart failure and at autopsy showed "almost total absence of the myocardium in the right ventricle in the absence of fatty tissue." "Examination of the cut edge of the ventricle wall reveals it to be paper-thin with no myocardium visible…."

The eponym Uhl's anomaly has been employed in adults with parchment RV. It is now also clear that the papyraceous appearance of the ventricular free wall is the end stage of an acquired, genetically determined progressive loss of myocardium, as in the Osler case [10].

The infant reported by Uhl was affected by a cardiac structural defect present at birth and, as such, falls into the category of congenital heart disease. Nonetheless, we cannot exclude that in infants with Uhl's anomaly the myocyte loss might have started during fetal life.

The history of the disease at our university started in the 1960s, when Professor Sergio Dalla Volta, the founder of modern cardiology in Padua with cardiac catheterization, published a series of cases featured haemodynamically by "auricularization of the right ventricular pressure" to underlie the absence of an effective systolic contraction of the RV, when a pressure curve was recorded in the RV similar to that of the right atrium, with the blood pushed from the right atrium directly to the pulmonary artery [11,12]. Thirty years later the heart of one of these patients was studied following cardiac transplantation at the age of 65, and had a parchment RV with an almost intact left ventricle [2].

In 1978, the late Professor Vito Terribile of our institute performed an autopsy of a woman with a history of palpitations and congestive heart failure, who died of pulmonary thromboembolism. The heart showed dilatation of the RV with mural thrombosis, "adipositas cordis" even at the posterior wall and apex (like the Laennec description), and "myocardial sclerosis of the left ventricle" in the absence of coronary artery disease, in keeping with what we now call biventricular arrhythmogenic cardiomyopathy.

The arrhythmic propensity of this substrate was first discovered in the 1970s by Guy Fontaine who demonstrated that life-threatening ventricular tachyarrhythmias with left bundle branch block morphology can originate from the RV [13]. The basal ECG may show delayed depolarization with an epsilon wave at the end of the QRS complex, which he named post-excitation syndrome. This differs from the pre-excitation syndrome (Wolff–Parkinson–White syndrome) with delta wave preceding the QRS complex.

In the 1980s, Marcus and Fontaine [14] reported a series of adult patients with this disease presenting with ventricular arrhythmias with left bundle branch block morphology. Microscopic examination of myocardial samples, removed at surgical disconnection of the right from the left ventricle, disclosed fibrofatty replacement that the authors interpreted as a maldevelopmental defect, and called the disease "right ventricular dysplasia." The term was then replaced by cardiomyopathy in the WHO nomenclature and classification of heart muscle disease [15].

The study of Marcus et al. was limited to adult patients and the ventricular arrhythmias of RV origin that were neither considered malignant nor interpreted as an inherited disease [14].

Meanwhile knowledge of the disease made progress in Padua, thanks to Andrea Nava, a true pioneer in the field of cardiovascular clinical genetics. He realized the genetic inheritance of the disease with a Mendelian dominant transmission, introducing the concept of "genetically determined cardiomyopathy," since he showed the onset of the phenotype in childhood [16,17].

Thiene et al. [3] first draw the attention to the malignant aspect of the disease, presenting in youths with sudden death, even as its first manifestation. By collecting and studying all the cases of juvenile sudden death (<35 years) occurring in the Veneto region, Italy (nearly 5 million inhabitants), they showed that ARVC/D is a leading cause of sudden death in young athletes (Fig. 2.2). The subjects had inverted T waves in right precordial leads and apparently benign premature ventricular beats with left bundle branch block morphology in the ECG, which is compulsory in Italy for sports eligibility. In other words, it was demonstrated that the

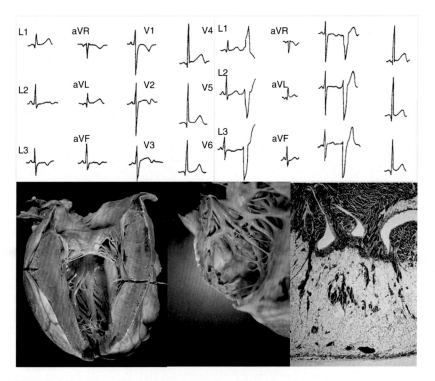

FIGURE 2.2 **A 30-year-old athlete who died suddenly during a soccer game and included in the original series published in 1988: note the inverted T waves in the right precordial leads.** At postmortem, the left ventricle was normal, whereas the right ventricle showed fibrofatty replacement of the free wall with inferior aneurysm. *Modified from Ref. [1].*

RV may be similar to the left ventricle in hypertrophic cardiomyopathy [18] and ischemic heart disease, and that ARVC/D may be a cause of premature death. Awareness of the heredofamilial nature of the disease stimulated molecular genetic investigation. By linkage analysis, the locus of a possible gene mutation was mapped to 14q23-q24 [19], oddly enough in the same chromosome of β-myosin heavy chain mutation in hypertrophic cardiomyopathy. At that time, there was no information on which gene might be the culprit, certainly neither sarcomeric genes since there was no evidence of hypertrophy and disarray, nor cytoskeleton and dystrophin complex as in dilated cardiomyopathy, since mechanical contractility was fairly preserved in the left ventricle. A revealing idea came from a group of Greek scholars from Naxos Island. In a letter to the editor, following our publication on ARVC/D and sudden death in the young, they drew the attention of the scientific community to a recessive cardiocutaneous syndrome, combining ARVC/D phenotype and palmoplantar keratosis and woolly hair [20,21]. Both epidermal and cardiac cells possess cell junction apparatus, ensuring mechanical adherences. Thus, protein genes of the desmosomal apparatus became candidates. Soon thereafter, a research group from Heidelberg demonstrated that knock-out transgenic mice for the JUP (plakoglobin gene), namely gamma catenin, resulted in severe myocardial injury with almost disappearance of desmosomes and spontaneous cardiac rupture during fetal life [22]. The authors said that "the human plakoglobin gene is located on chromosome 17q21, a region not yet identified in human cardiomyopathy patients."

As a consequence, JUP immediately became the candidate gene for Naxos disease. Linkage analysis was carried out in Naxos families, identifying the gene defect in locus 17q21 [23]. Subsequent gene sequencing proved that the molecular defect consists of a deletion of the JUP gene [24]. At about the same time, a dermatologist from Ecuador, Luis Carvajal Huerta, described another recessive cardiocutaneous syndrome, consisting of biventricular cardiomyopathy, woolly hair and palmoplantar keratosis quite similar to Naxos disease, with pump failure and rare polymorphic ventricular arrhythmias [25]. A mutation was demonstrated in a gene, encoding a major protein of the desmosomal apparatus, namely desmoplakin (DSP) [26]. One child of the original family reported by Carvajal died with congestive heart failure and the wife of Carvajal, sent the heart specimen to Dr. Saffitz in St. Louis [27]. Gaetano Thiene was asked to review the heart and found that it had a right ventricle fibrous replacement and typical aneurysms in the "triangle of dysplasia." The left ventricle was dilated and had a mural thrombus consistent with biventricular arrhythmogenic cardiomyopathy.

In Padua, we were looking for a candidate gene for the dominant variant of ARVC/D, pointed to DSP and a missense mutation was identified in some of Nava's families [28]. Genotype–phenotype correlations demonstrated that DSP proteins frequently presented with biventricular or even predominant left ventricular involvement [29].

Subsequently, other genes encoding desmosomal proteins were investigated in nonsyndromic ARVC/D patients and missense mutations were found in plakophilin-2 (PKP2), desmoglein-2 (DSG2), and desmocollin-2 (DSC2) [30–33]. Eventually both recessive and dominant variants of ARVC/D were found to be related to cell junction defects, so that the disease could be labeled as a desmosomal disease [34–36] (Fig. 2.3).

Basso et al. revealed ultrastructural abnormalities of the desmosome in the myocardium of genotyped ARVC/D patients, and suggested that disruption of the intercalated disc might be the final common pathway of a genetically determined myocyte death, resulting in fibrofatty scarring and arrhythmogenicity [37].

Nondesmosomal genes, such as transforming growth factor β (TGF-β3), transmembrane protein 43 (TMEM43), alpha-T catenin (CTNNA3), titin (TNT), desmin (DES), PLN (phospholamban), and LMNA (lamin A/C), have also been found [38–44].

Ryanodine receptor 2 (RyR2), originally reported as a form of ARVC/D, was eventually related to a distinct morbid entity (catecholaminergic polymorphic ventricular tachycardia), an ion channel disease without substrate [45].

Transgenic mice overexpressing the mouse homologous of desmosomal gene mutation were generated [46–49]. In the desmoglein transgenic mouse, published by Pilichou et al. [49], the recapitulated disease consists of dilatation and aneurysm of both ventricles with fibrous replacement of the myocardium, prolonged electrical epicardial ventricular activation at electrophysiology, tachyarrhythmias, and even sudden death. It was then shown by Rizzo et al. [50] that delayed electrical ventricular activation and arrhythmia inducibility occurs well before the onset of myocyte death and scarring, as a consequence of Na++ current density. This is most probably due to cross talk between cell junction and sodium channel complex, consistent with the existence of an electrical disturbance even in the early stages of ARVC/D.

The discovery of ARVC/D as a distinct biological nosographic entity, with a specific genetic background, was paralleled by advances in diagnosis and treatment.

Diagnostic criteria have been published in 1994 based on noninvasive (ECG, echo) and invasive (angiography, endomyocardial biopsy) studies [51] (Fig. 2.4). In 2010, diagnostic criteria were updated including quantitative parameters [52], even for endomyocardial biopsy [53].

Cardiac magnetic resonance (CMR) was found to be an excellent diagnostic procedure, to detect both morphofunctional (poor contractility, ventricular dilatation, aneurysms, and dyskinesia) and tissue

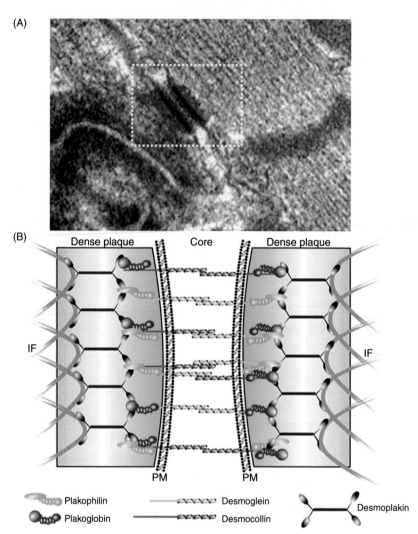

FIGURE 2.3 **Intercellular mechanical junction (desmosome) of the cardiomyocyte.**
(A) Transmission electron microscopy of cardiomyocyte desmosome (boxed area)×80.000.
(B) Schematic representation of the desmosome components. It consists of a core region,
which mediates cell–cell adhesion, and a dense plaque, which provides attachment to the
intermediate filaments. There are three major groups of desmosomal proteins: (1) transmembrane proteins (i.e., desmosomal cadherins) including desmocollin and desmoglein;
(2) Desmoplakin (DSP), a plakin family protein that binds directly to intermediate filaments
(desmin in the heart); and (3) linker proteins (i.e., armadillo family proteins) including
plakoglobin and plakophilin, which mediate interaction between the desmosomal cadherin
tails and DSP. IF, Intermediate filaments; PM, plasma membrane. *Modified from Ref. [35].*

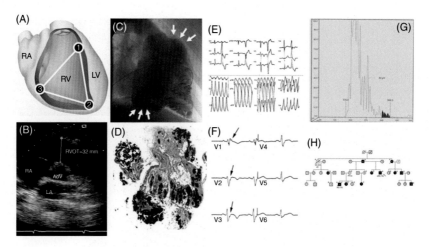

FIGURE 2.4   **Diagnostic morphofunctional, electrocardiographic, and tissue character-istic features of ARVC/D.** (A) Diagram of the "triangle of dysplasia," which illustrates the characteristic areas for structural and functional abnormalities of the RV (LV, left ventricle; RA, right atrium; RV, right ventricle); (B) 2D echocardiography showing RV outflow tract en-largement from the parasternal short-axis view (AoV, aortic valve; LA, left atrium; RA, right atrium; RVOT, right ventricle outflow tract); (C) RV contrast angiography (30° right anterior oblique view) demonstrating localized RVOT as well as inferobasal aneurysms (arrows) with mild tricuspid regurgitation; (D) endomyocardial biopsy sample with extensive myocardial atrophy and fibrofatty replacement (trichrome; ×6); (E) 12-lead ECG with inverted T waves (V_1, V_2,V_3) with left bundle branch block (LBBB) morphology premature ventricular beats and ventricular tachycardia (VT); (F) ECG tracing showing postexcitation epsilon wave in precordial leads V_1,V_2, V_3 (arrows); (G) signal-averaged ECG with late potentials (40-Hz high-pass filtering); filtered QRS duration (QRS), 217 ms; low amplitude signal (LAS), 107 ms, and root-mean-square voltage of terminal 40 ms (RMS), 4 μV; (H) family pedigree of ARVC/D: note the autosomal dominant inheritance of the disease with 50% of offspring affected. *Modified from Ref. [35].*

characterization, for both fatty tissue and fibrosis by the late enhancement technique. The use of CMR with gadolinium is extremely effective in de-tecting left ventricular involvement [54] (Figs 2.5 and 2.6).

Electroanatomic mapping (EM) was able to reveal "electric scars," namely areas of reduced or absent electrical activity, equivalent to fibro-fatty myocardial atrophy. The correlation with endomyocardial biopsy and MRI was important for the differential diagnosis with nonischemic arrhythmogenic diseases like sarcoidosis, myocarditis, RV outflow tract tachycardia, and Brugada syndrome [55–58].

Meanwhile, great advances have been achieved in the prevention of premature death in ARVC/D. Lifestyle and sports disqualification, by detecting ECG-specific alterations (T wave inversion in right precordial leads, QRS widening, and epsilon postexcitation wave) can be lifesav-ing, considering that physical activity is the main precipitating factor of

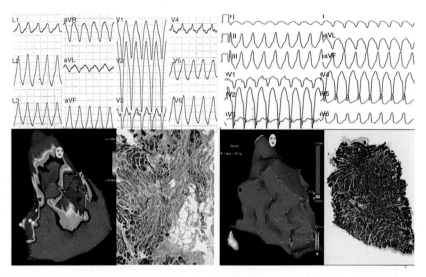

FIGURE 2.5   Electroanatomic mapping is a fundamental tool in the differential diagnosis between segmental infundibular arrhythmogenic right ventricular cardiomyopathy (left panel) and idiopathic right ventricular outflow tract tachycardia (right panel). *Modified from Ref. [1].*

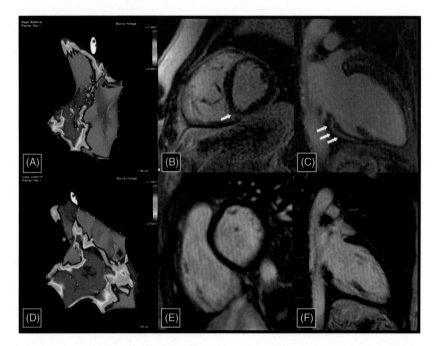

FIGURE 2.6   Electroanatomic mapping (EM) is more sensitive than cardiac magnetic resonance with late enhancement to detect right ventricular involvement in arrhythmogenic right ventricular cardiomyopathy. However, the left ventricle is frequently involved, and may be considered the 'mirror' of the right ventricle on cardiac magnetic resonance. *Modified from Ref. [1].*

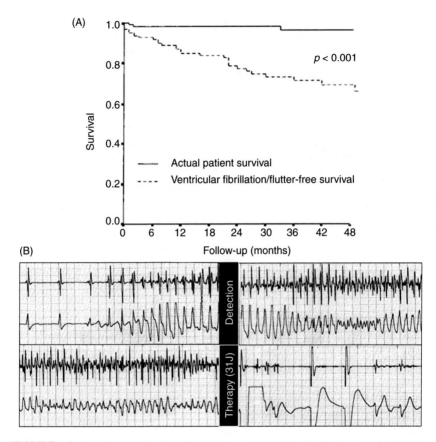

FIGURE 2.7 **ICD therapy in ARVC/D.** (A) Projected survival of 132 patients with ARVC/D who had implanted cardiac defibrillator (ICD). The Kaplan–Meier analysis compares actual patient survival (continuous line) with survival free from either VF or ventricular flutter (dotted line) that would have been probably fatal in the absence of an appropriate ICD intervention. The divergence between curves reflects the estimated survival benefit conferred by ICD therapy. At 36 months, actual total patient survival was 96%, compared with 72% VF and ventricular flutter survival. (B) Stored intracardiac electrogram in an ARVC/D patient. This shows one episode of VF, with appropriate detection and successful ICD discharge followed by sinus rhythm. *Modified from Ref. [35].*

cardiac arrest [59–61]. Drug therapy and ablation, although palliative measures, are of help [62–64]. A cardiac defibrillator, either implantable or external, can revert cardiac arrest, due to ventricular fibrillation [65,66] (Fig. 2.7).

However, cure rather than palliation of ARVC/D should be pursued, by intervening in the pathobiology of the disease, namely the onset and progression of cell death leading to myocardial dystrophy. Arrhythmias should not be the only target in patients with ARVC/D to prevent sudden death [62] (Fig. 2.8).

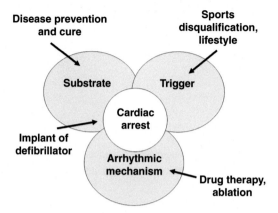

FIGURE 2.8 **Cardiac arrest is the combination of trigger, substrate, and arrhythmic mechanism.** Sport acts as a trigger; drug therapy and ablation are palliative measures. Defibrillator clearly intervenes on the mode of cardiac arrest, namely ventricular fibrillation. The true disease prevention and cure should point to onset and progression of the substrate.

Cellular reprogramming of somatic cells into pluripotent stem cells (iPSCs) enables patient-specific *in vitro* remodeling of human genetic disorders for pathogenetic investigation and drug screening [67–69]. Fibroblasts of patients affected by ARVC/D may be used to generate autologous cardiomyocytes. Zebrafish model may also be employed in the study of ARVC/D to elucidate pathogenetic mechanisms and screen drug therapy [70].

Engineering adeno-associated viral vectors containing c-DNA of wild-type desmosomal genes, which may be transferred into the heart, may represent a curative gene therapy [71].

We are entering the era of molecular medicine, and the time has come for myocardial dystrophy (ARVC/D) as it has been for muscular dystrophy (Duchenne) [72].

PATHOLOGY AND ENDOMYOCARDIAL BIOPSY

The pathological hallmark for the diagnosis of ARVC/D is based on gross and histologic evidence of transmural fibrofatty replacement of the myocardial free wall [3,73]. It is a wave-front phenomenon from the epicardium to the endocardium, sparing the trabeculae that sometimes may appear hypertrophic, mimicking noncompacted myocardium.

The atrophy of the myocardium results in aneurysms, located at the apex, infundibulum, and posteroinferior wall (triangle of dysplasia). The latter (subtricuspid) may be considered a pathognomonic (Fig. 2.9) feature of the disease. At gross examination, the right side of the heart appears yellowish or whitish because of fibrofatty replacement (Figs 2.9 and 2.10).

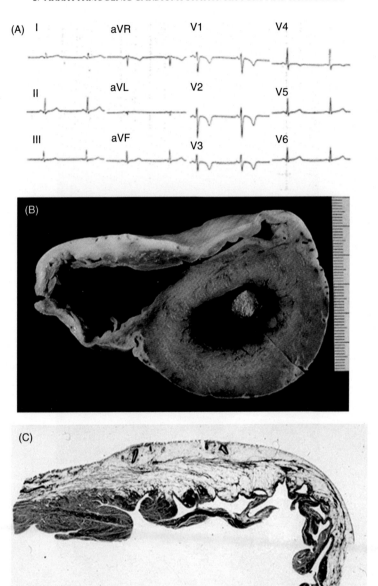

FIGURE 2.9 **A 14-year-old boy died suddenly during a soccer game:** note the inverted T waves in the right precordial leads with isolated premature ventricular beat and LBBB morphology (A). At postmortem gross (B), and histological (C) examination, fibrofatty replacement of the infundibular RV free wall was found (B). (Heidenhain trichrome, note the inferior aneurysm) *Modified from Ref. [2].*

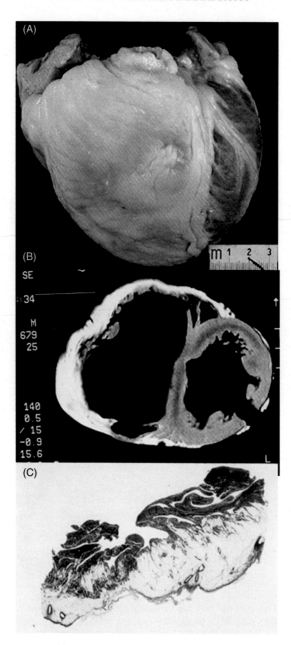

FIGURE 2.10 **A 49-year-old woman who had a cardiac transplant due to heart failure and refractory ventricular arrhythmias.** (A) External view of the native heart specimen obtained at cardiac transplantation: note the yellow appearance of the right side of the heart; (B) *in vitro* spin-echo CMR, short axis, shows massive RV dilatation and full-thickness myocardial atrophy with high intensity signal; (C) panoramic histologic section of the RV free wall shows transmural fibrofatty replacement (Heidenhain trichrome). *Modified from Ref. [2].*

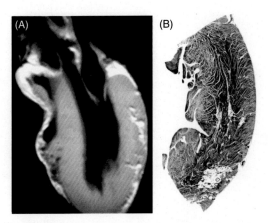

FIGURE 2.11   **A 14-year-old boy died suddenly during a soccer game.** At postmortem, including in vitro CMR (A), left ventricular involvement with fibrofatty replacement was also observed (B) (same case as Fig. 2.9).

The heart weight, particularly in case of sudden death in the young, is normal or at the upper limits of normal (350–400 g), with moderate to severe RV enlargement. The heart removed at the time of transplantation because of congestive heart failure may show cardiomegaly (up to 600 g) with biventricular involvement. In the setting of congestive heart failure, mural thrombosis may be observed in the ventricles and, in the presence of atrial fibrillation, in right and/or left atrial appendage. They may be the source of pulmonary or cerebral-thromboembolism with stroke. Thickening of the endocardium may occur due to thrombus deposition and organization.

The pathologic process may be diffuse or, more rarely, segmental with isolated involvement of the infundibulum, apex, or inferior wall.

Focal left ventricular involvement is observed in nearly 70% of cases [73] (Fig. 2.11). Recent observations from athletes with sudden death show as a form with isolated left ventricular involvement and posterolateral subepicardial fibrofatty scar, which may not show by ECG. The only identification *in vivo* is by CMR [54, 74].

Involvement of the ventricular septum is much less frequent (20%) suggesting that the disease involves the subepicardial (free walls) rather than subendocardial (septum).

The amount of fatty and fibrous tissue, replacing the myocardium, is variable. There are cases with prevalent fatty infiltration ("lipomatous variant") and cases with prevalent fibrous replacement ("fibrous variant") [3,73] (Fig. 2.12). The "lipomatous variant" may have increased thickness of the RV free wall (so-called pseudohypertrophy); nonetheless a small amount of replacement-type fibrous tissue, at least

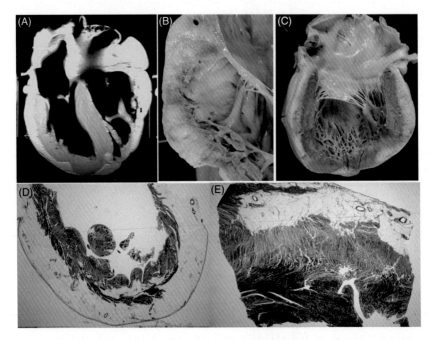

FIGURE 2.12   **A 40-year-old man, previously asymptomatic, who died suddenly at rest.** (A) *In vitro* spin-echo CMR, four-chamber cut, shows increased high intensity signal in both ventricles, either transmural (right) or subepicardial (left); (B) view of the RV with fatty appearance of the lateral wall, subtricuspid aneurysm, and endocardial fibrous thickening; (C) view of the posterolateral left ventricular free wall: note the wave-front extension of fat from the epicardium toward the endocardium; (D) panoramic histologic section of the RV free wall shows transmural fibrofatty replacement (Heidenhain trichrome); (E) panoramic histologic section of the left ventricular free wall with fibrofatty replacement in the outer layer (Heidenhain trichrome). *Modified from Ref. [2].*

in the subendocardial layer, is always seen at histology with collagen staining [73,75].

The fibrosis of the fibrofatty variant is usually associated with thinning of the ventricular free wall, which appears parchment like and translucent. This accounts for the formation of aneurysms.

Myocyte death or degeneration is observed microscopically, associated with sign of adipogenesis, all consistent with the concept of myocyte injury and repair. Inflammatory infiltrates are almost a regular finding [73,76] (Fig. 2.13). An inflammatory pathogenesis has been postulated, even though a phlogistic reaction to spontaneous cell death is a more plausible explanation. The role of viruses in the etiopathogenesis of the disease has been excluded by molecular investigations [77]. Inflammatory reaction, whether primary or secondary, may act as a trigger of abrupt electrical instability and arrhythmic death.

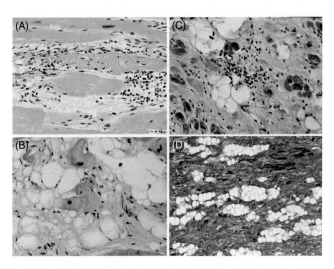

FIGURE 2.13   **Histologic features of ARVC/D.** (A) Contraction band myocyte necrosis and mononuclear inflammatory infiltrates (Hematoxylin–eosin); (B) myocytolysis with early adipocytes and fibroblasts infiltration (Hematoxylin–eosin); (C) mature fibrous tissue and fatty tissue with residual inflammatory reaction (Hematoxylin–eosin); (D) islands of surviving myocytes are entrapped within fibrous and fatty tissue (Heidenhain trichrome). *Modified from Ref. [2].*

The basic phenomenon is progressive cell death, which may be patchy through a mechanical breakdown of the cardiomyocytes as a result of wall stretching during effort or due to apoptosis/necroptosis [78]. An infarct-like onset of cardiomyocyte death, located in the subepicardium, has been shown to occur, especially in the left ventricle and explains cases of isolated left ventricular involvement (left ventricular arrhythmogenic cardiomyopathy) [79].

Fatty infiltration of the right ventricle "per se" should not be regarded as a hallmark of ARVC/D. The original description of a lipomatous variant has been a source of misleading diagnoses, since it has not been sufficiently appreciated that even in the lipomatous variant a certain amount of replacement-type fibrosis should be found to label the diagnosis as ARVC/D. Fat is a normal finding in the outer layer of the RV free wall, particularly in the anterior wall and at the apex. Eighty-five percent of the hearts from people who died from extracardiac causes contain some myocardial fatty infiltration of the RV, especially in older female subjects. The myocytes appear to be pushed apart rather than replaced, without any evidence of fibrosis, myocyte degeneration, or inflammation [75] (Fig. 2.14). Fatty infiltration implies thickening of the RV free wall, without aneurysms, which is a pathognomonic feature of ARVC/D. Moreover, increase of fatty tissue in the subepicardium is a regular finding in the obese (adipositas cordis) and should not be interpreted as ARVC/D.

In a patient with sudden death, when extensive fatty infiltration is observed in the RV free wall, it would be preferable to report the finding

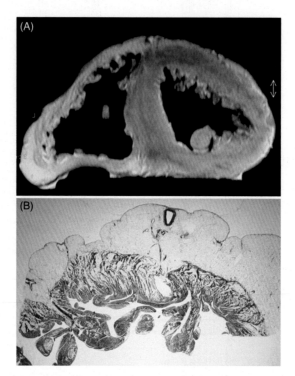

FIGURE 2.14 **Adipositas cordis in an obese individual.** (A) *In vitro* spin-echo CMR, short-axis cut, shows transmural high intensity signal in the right ventricle, with preserved wall thickness; (B) panoramic histologic section of the RV free wall shows increased epicardial fat and finger-like fat infiltration of the underlying myocardium, in the absence of fibrous tissue (Heidenhain trichrome). *Modified from Ref. [2].*

without any implication in term of cause and effect relationship, otherwise the risk is a misdiagnosis [80].

A forensic autopsy investigation on sudden death in France reported a high prevalence (>10%) of ARVC/D in sudden death cases aged 1–65 years, without any evidence of fibrous tissue replacement, a finding clearly not tenable [81]. In fact, in ARVC/D hearts of patients who died suddenly, with proven mutation of genes encoding desmosomal proteins, the hearts are typically characterized by biventricular involvement, aneurysms, fibrofatty replacement, myocyte death, and nuclear abnormalities (unpublished data).

ARVC/D has been originally reported as a disorder predominantly affecting the RV. Since then, left ventricular involvement has been recognized with greater frequency. In pathology series, the incidence of left ventricular involvement may be up to 70% [73,82]. The incidence increases up to 87% in heart specimens from cardiac transplantation and by examination of histological slides from multiple blocks of the left ventricle and ventricular septum [83].

Full-thickness left ventricular transmural fibrofatty infiltration with aneurysm formation is rarely reported in the literature on ARVC/D. The lesion is typically subepicardial or midmural, with greater amount of fibrosis as compared to the RV. A wave-front of subepicardial fatty infiltration is also a hallmark of left ventricular involvement (Figs 2.15 and 2.16). "Pure" left ventricular involvement without or with minimal evidence of RV involvement has been reported, particularly in athletes who escaped detection of the disease at the preparticipation screening with ECG. Genotype–phenotype correlations pointed to DSP as a causative gene of "arrhythmogenic left ventricular cardiomyopathy." Aguilera et al. reported biventricular disease in 62%, or isolated right or left ventricular involvement each in 19% [84]. These data suggest that the various locations and amount of fibrofatty tissue are different expressions of the same genetic disease (one gene – different phenotype). Therefore, consideration should be given to employ the term arrhythmogenic cardiomyopathy (ruling out right, left, or biventricular) or, desmosomal cardiomyopathy or myocardial dystrophy.

In vivo, the histopathologic diagnosis of ARVC/D is feasible by endomyocardial biopsy of the RV, since the fibrofatty replacement is usually transmural, thus detectable on the endomyocardial approach [53,85]. The left ventricular approach from the retrograde aorta is not advisable, because the pathologic substrate, when present, is subepicardial and not reachable by the endocardial bioptome. Moreover, the fibrofatty phenomenon is rarely located in the ventricular septum, then the bioptome should point to the free wall and/or adjacent region. When adipose tissue includes nerves and mesothelial cells, this clearly indicates an epicardial source. Moreover, this may be a sign that the bioptome perforated the RV free wall. This complication rarely accounts for cardiac tamponade, and is associated with bleeding and heals spontaneously. The definitive diagnosis of ARVC/D relies on the histological presence of myocardial atrophy with fibrofatty replacement of the RV myocardium and is listed among the Task Force criteria [51,52]. The early 1994 Task Force Diagnostic Criteria were based only on a qualitative analysis of endomyocardial biopsy samples (presence or not of fibrofatty replacement) [51]. However, the presence of fatty tissue in the RV free wall is neither specific nor pathognomonic of ARVC/D. It is observed in the elderly or in obese people and fibrous tissue is present in other cardiomyopathies. The key to diagnosis is the quantity rather than the quality of myocardial tissue replacement. Morphometric criteria have been put forward for ARVC/D diagnosis and included in the 2010 version of diagnostic criteria [52,53]. A residual amount of myocardium (less than 60%), caused by fibrous or fibrofatty replacement, has been calculated in *in vitro* specimens to have a high diagnostic accuracy and now is listed among the major criteria for the diagnosis of ARVC/D (Fig. 2.17). The

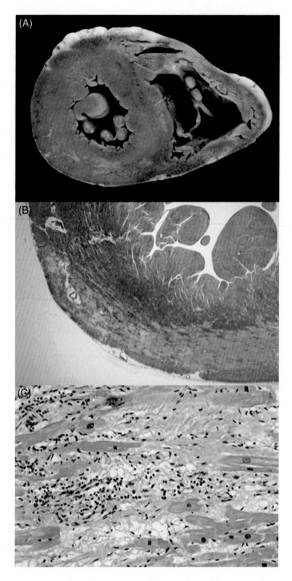

FIGURE 2.15 **A 15-year-old boy, family member of a proband with ARVC/D due to DSP mutation, who died suddenly at rest despite negative cardiological screening.** (A) Cross section of the heart: there is no macroscopic evidence of fatty tissue infiltration nor aneurysm in the right ventricle, whereas a gray band is evident in the subepicardial posterolateral region of the left ventricle; (B) panoramic histologic view of the left ventricular wall showing a subepicardial band of acute–subacute myocyte necrosis with loose fibrous tissue and granulation tissue (trichrome Heidenhain); (C) myocyte necrosis, myocytolysis, and polymorphous inflammatory infiltrates together with fibrous and fatty tissue repair are visible at higher magnification of B (Hematoxylin–eosin). *Modified from Ref. [2].*

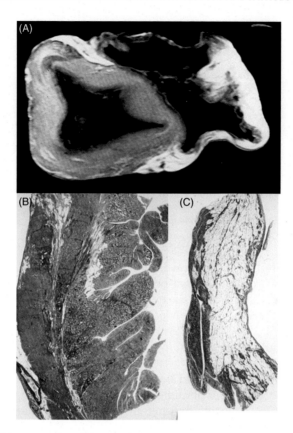

FIGURE 2.16 **A 35-year-old man who died suddenly with a clinical diagnosis of ARVC/D with biventricular involvement, belonging to a family found to have a DSP mutation at genetic screening.** (A) *In vitro* spin-echo CMR, short-axis cut, shows biventricular dilatation, transmural fatty infiltration of the RV free wall and spots of fatty tissue in the posterolateral left ventricular free wall; (B) panoramic histologic section of the left ventricular lateral wall: fibrofatty with prevalent fibrous tissue replacement is evident in the subpericardial and midmural layers (Heidenhain trichrome); (C) panoramic histologic section of the RV free wall transmural fibrofatty replacement of the myocardium (Heidenhain trichrome). *Modified from Ref. [2].*

diagnostic sensitivity may be improved if the bioptome is guided by either EM or CMR imaging.

Moreover, endomyocardial biopsy is essential for differential diagnosis with the so-called phenocopies, such as myocarditis, sarcoidosis, or idiopathic outflow tract tachycardia [55,56,58]. Idiopathic RV outflow tract tachycardia, a benign nonfamilial disease, has normal myocardium at endomyocardial biopsy and no electric scar at the EM [56]. Myocarditis, a sporadic transmissible disease mostly due to viruses,

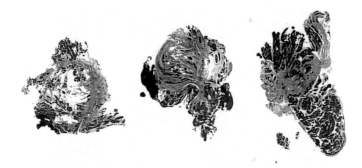

FIGURE 2.17 **Endomyocardial biopsy findings in a proband affected by a diffuse form of ARVC/D: all three bioptic samples show extensive fibrofatty tissue replacement (Heidenhain trichrome).** *Modified from Ref. [2].*

may present with RV tachycardia, scarring at EM and histologic evidence of myocardial inflammatory infiltrates [55]. Sarcoidosis is quite intriguing and challenging, since the ECG and imaging may be highly suggestive of ARVC/D, but histology may reveal noncaseous granulomas with giant cells [58,86]. Positron emission tomography may show lymph node or other organ locations, consistent with extracardiac sarcoidosis.

Quantitative criteria should not exclude qualitative evaluation of the biopsy microscopically. Replacement-type fibrosis including some inflammatory infiltrates, myocyte degeneration, and evidence of adipogenesis are microscopic hallmark of ARVC/D.

Following the discovery that familial ARVC/D is due to mutations of genes encoding desmosomal proteins, it has been postulated that reduced signaling of junctional proteins such as plakoglobin may be diagnostically specific at tissue evaluation [87]. Defective signal of plakoglobin from living myocyte was shown to exist in biopsy specimen of patients affected by ARVC/D (Fig. 2.18). It might assist in establishing the diagnosis, not on the basis of the amount of atrophic myocardial tissue (fibrofatty replacement), but from the living cardiomyocytes itself. This facilitates the value of endomyocardial biopsy, by retrieving myocardial tissue from the ventricular septum, thus avoiding the free wall at risk of perforation. Unfortunately, the findings are not specific, since a defective plakoglobin signal occurs in other nondesmosomal cardiomyopathies and inflammatory myocardial conditions [88].

ARVC/D phenotype has been recently recapitulated in transgenic animal models [46–49]. This provides the opportunity to review the morphologic features of the disease in these models. The study of a transgenic mouse model with cardiac overexpression of DSG2 gene mutation N271S (Tg-NS) yielded important information into the pathobiologic

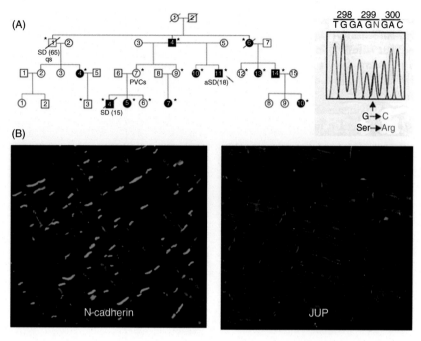

FIGURE 2.18   **Immunoreactive plakoglobin signal and histologic features in a sudden death victim from familial ARVC/D caused by a mutant DSP gene (the same case as Fig. 2.15).** (A) Family pedigree with identified mutation (S299R) in exon 7 of DSP gene; (B) immunohistochemical analysis of human myocardial samples of the proband who died suddenly at the age of 15 years shows a marked reduction in immunoreactive signal levels for plakoglobin (JUP) (right) but normal signal levels for the nondesmosomal N-cadherin adhesion molecule (left).

mechanisms involved in the onset and progression of ARVC/D [49]. The clinical features of human ARVC/D were reproduced, including spontaneous ventricular arrhythmias, cardiac contractility dysfunction, biventricular dilatation with aneurysms, and early arrhythmic sudden death. The study at histology and transmission electron microscopy showed that myocyte necrosis initiates myocardial loss. Electron microscopy in Tg-NS mice, aged 2–3 weeks, showed disruption of sarcolemma and mitochondria swelling with disintegration of myofilaments and cytoplasmic organelles, consistent with myocyte necrosis. Cardiomyocyte cell death triggers inflammatory reaction and eventually calcification, followed by fibrous repair and aneurysm formation (Fig. 2.19). This course of pathologic injury and repair is consistent with the concept of arrhythmogenic cardiomyopathy as a genetically determined heart muscle disease. Curative molecular therapy should be directed to prevent the onset and/or slow the progression of cardiomyocyte death.

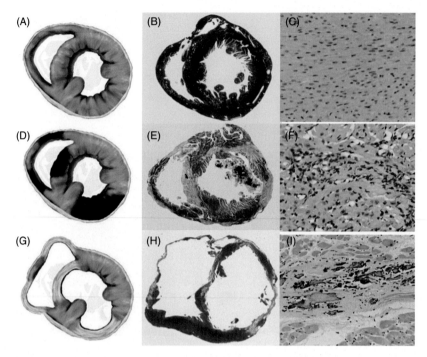

FIGURE 2.19 **Pathology of ARVC/D in transgenic animal models with desmoglein2 mutation overexpression.** The disease consists of a progressive myocardial dystrophy (genetically determined cardiomyopathy), with a normal heart at birth (A, B, C), followed by myocardial necrosis with a wave-front extension from the epicardium toward the endocardium (D, E, F), eventually leading to biventricular aneurysms due to wall thinning and calcification (G, H, I).

References

[1] Thiene G. The research venture in arrhythmogenic right ventricular cardiomyopathy: a paradigm of translational medicine. Eur Heart J 2015;36:837–48.

[2] Marcus FI, Nava A, Thiene G. Arrhythmogenic right ventricular cardiomyopathy/dysplasia – recent advances. Milano: Springer; 2007.

[3] Thiene G, Nava A, Corrado D, Rossi L, Pennelli N. Right ventricular cardiomyopathy and sudden death in young people. N Engl J Med 1988;318:129–33.

[4] Lancisi GM. De motu cordis et aneurysmatibus. Caput V. Naples: Excudebat Felix-Carolus Musca; 1736.

[5] Laennec RTH. A treatise on the diseases of the chest and on mediate auscultation. Paris: Brosson & Chaudé; 1819.

[6] Elliot G. Middlemarch. 1st ed. Edinburgh and London: William Blackwood and Sons; 1871.

[7] Osler W. The principles and practice of medicine. 6th ed. New York: D. Appleton & Co; 1905. p. 20.

[8] Segall HN. Parchment heart (Osler). Am Heart J 1950;40:948–50.

[9] Uhl HSM. A previously undescribed congenital malformation of the heart: almost total absence of the myocardium of the right ventricle. Bull Johns Hopkins Hosp 1952;91: 197–209.

[10] Basso C, Corrado D, Thiene G. Arrhythmogenic right ventricular cardiomyopathy: what's in a name? From a congenital defect (dysplasia) to a genetically determined cardiomyopathy (dystrophy). Am J Cardiol 2010;106:275–7.

[11] Dalla Volta S, Battaglia G, Zerbini E. Auricularization of right ventricular pressure curve. Am Heart J 1961;61:25–33.

[12] Dalla Volta S, Fameli O, Maschio G. Le sindrome clinique et hemodynamique de l'auricularisation du ventricule droit. Arch Mal Cœur 1965;58:1129–43.

[13] Fontaine G, Frank R, Gallais-Hamonno F, Allali I, Phan-Thuc H, Grosgogeat Y. Electrocardiography of delayed potentials in post-excitation syndrome. Arch Mal Coeur Vaiss 1978;71:854–64.

[14] Marcus FI, Fontaine G, Guiraudon G, Frank R, Laurenceau JL, Malergue S, Grosgogeat Y. Right ventricular dysplasia. A report of 24 adult cases. Circulation 1982;65:384–98.

[15] Richardson P, McKenna WJ, Bristow M, Maisch B, Mautner B, O'Connell J, Olsen O, Thiene G, Goodwin J, Gyarfas I, Martin I, Nordet P. Report of the 1995 WHO/ISFC Task Force on the definition and classification of cardiomyopathies. Circulation 1996;93:841–2.

[16] Nava A, Thiene G, Canciani B, Scognamiglio R, Daliento L, Buja G, Martini B, Stritoni P, Fasoli G. Familial occurrence of right ventricular dysplasia: a study involving nine families. J Am Coll Cardiol 1988;12:1222–8.

[17] Nava A, Bauce B, Basso C, Muriago M, Rampazzo A, Villanova C, Daliento L, Buja G, Corrado D, Danieli GA, Thiene G. Clinical profile and long-term follow-up of 37 families with arrhythmogenic right ventricular cardiomyopathy. J Am Coll Cardiol 2000;36: 2226–33.

[18] Maron BJ. Right ventricular cardiomyopathy. Another cause of sudden death in the young. N Engl J Med 1988;318:178–9. (Editorial).

[19] Rampazzo A, Nava A, Danieli GA, Buja G, Daliento L, Fasoli G, Scognamiglio R, Corrado D, Thiene G. The gene for arrhythmogenic right ventricular cardiomyopathy maps to chromosome 14q23-q24. Hum Mol Genet 1994;3:959–62.

[20] Protonotarios N, Tsatsopoulou A, Scampardonis G. Familial arrhythmogenic right ventricular dysplasia associated with palmoplantar keratosis. N Engl J Med 1988;319:174–6.

[21] Protonotarios N, Tsatsopoulou A, Patsourakos P, Alexopoulous D, Gezerlis P, Simitsis S, Scampardonis G. Cardiac abnormalities in familial palmoplantar keratosis. Br Heart J 1986;56:321–6.

[22] Ruiz P, Brinkmann V, Ledermann B, Behrend M, Grund C, Thalhammer C, Vogel F, Birchmeier C, Günthert U, Franke WW, Birchmeier W. Targeted mutation of plakoglobin in mice reveals essential functions of desmosomes in the embryonic heart. J Cell Biol 1996;135:215–25.

[23] Coonar AS, Protonotarios N, Tsatsopoulou A, Needham EW, Houlston RS, Cliff S, Otter MI, Murday VA, Mattu RK, McKenna WJ. Gene for arrhythmogenic right ventricular cardiomyopathy with diffuse nonepidermolytic palmoplantar keratoderma and woolly hair (Naxos disease) maps to 17q21. Circulation 1998;97:2049–58.

[24] McKoy G, Protonotarios N, Crosby A, Tsatsopoulou A, Anastasakis A, Coonar A, Norman M, Baboonian C, Jeffrey S, McKenna WJ. Identification of a deletion in plakoglobin in arrhythmogenic right ventricular cardiomyopathy with palmoplantar keratoderma and woolly hair (Naxos disease). Lancet 2000;355:2119–24.

[25] Carvajal-Huerta L. Epidermolytic palmoplantar keratoderma with woolly hair and dilated cardiomyopathy. J Am Acad Dermatol 1998;39:418–21.

[26] Norgett EE, Hatsell SJ, Carvajal-Huerta L, Cabezas JC, Common J, Purkis PE, Whittock N, Leigh IM, Stevens HP, Kelsell DP. Recessive mutation in desmoplakin disrupts

desmoplakin-intermediate filament interactions and causes dilated cardiomyopathy, woolly hair and keratoderma. Hum Mol Genet 2000;9:2761–6.

[27] Kaplan SR, Gard JJ, Carvajal-Huerta L, Ruiz-Cabezas JC, Thiene G, Saffitz JE. Structural and molecular pathology of the heart in Carvajal syndrome. Cardiovasc Pathol 2004;13:26–32.

[28] Rampazzo A, Nava A, Malacrida S, Beffagna G, Bauce B, Rossi V, Zimbello R, Simionati B, Basso C, Thiene G, Towbin JA, Danieli GA. Mutation in human desmoplakin domain binding to plakoglobin causes a dominant form of arrhythmogenic right ventricular cardiomyopathy. Am J Hum Genet 2002;71:1200–6.

[29] Bauce B, Basso C, Rampazzo A, Beffagna G, Daliento L, Frigo G, Malacrida S, Settimo L, Danieli G, Thiene G, Nava A. Clinical profile of four families with arrhythmogenic right ventricular cardiomyopathy caused by dominant desmoplakin mutations. Eur Heart J 2005;26:1666–75.

[30] Gerull B, Heuser A, Wichter T, Paul M, Basson CT, McDermott DA, Lerman BB, Markowitz SM, Ellinor PT, MacRae CA, Peters S, Grossmann KS, Michely B, Sasse-Klaassen S, Birchmeier W, Dietz R, Breithardt G, Schulze-Bahr E, Thierfelder L. Mutations in the desmosomal protein plakophilin-2 are common in arrhythmogenic right ventricular cardiomyopathy. Nat Genet 2004;36:1162–4.

[31] Pilichou K, Nava A, Basso C, Beffagna G, Bauce B, Lorenzon A, Frigo G, Vettori A, Valente M, Towbin J, Thiene G, Danieli GA, Rampazzo A. Mutations in desmoglein-2 gene are associated with arrhythmogenic right ventricular cardiomyopathy. Circulation 2006;113:1171–9.

[32] Syrris P, Ward D, Evans A, Asimaki A, Gandjbakhch E, Sen-Chowdhry S, McKenna WJ. Arrhythmogenic right ventricular dysplasia/cardiomyopathy associated with mutations in the desmosomal gene desmocollin-2. Am J Hum Genet 2006;79:978–84.

[33] Beffagna G, De Bortoli M, Nava A, Salamon M, Lorenzon A, Zaccolo M, Mancuso L, Sigalotti L, Bauce B, Occhi G, Basso C, Lanfranchi G, Towbin JA, Thiene G, Danieli GA, Rampazzo A. Missense mutations in desmocollin-2 N-terminus, associated with arrhythmogenic right ventricular cardiomyopathy, affect intracellular localization of desmocollin-2 *in vitro*. BMC Med Genet 2007;8:65.

[34] Thiene G, Corrado D, Basso C. Cardiomyopathies: is it time for a molecular classification? Eur Heart J 2004;25:1772–5.

[35] Basso C, Corrado D, Marcus FI, Nava A, Thiene G. Arrhythmogenic right ventricular cardiomyopathy. Lancet 2009;373:1289–300.

[36] Basso C, Bauce B, Corrado D, Thiene G. Pathophysiology of arrhythmogenic cardiomyopathy. Nat Rev Cardiol 2011;9:223–33.

[37] Basso C, Czarnowska E, Della Barbera M, Bauce B, Beffagna G, Wlodarska EK, Pilichou K, Ramondo A, Lorenzon A, Wozniek O, Corrado D, Daliento L, Danieli GA, Valente M, Nava A, Thiene G, Rampazzo A. Ultrastructural evidence of intercalated disc remodelling in arrhythmogenic right ventricular cardiomyopathy: an electron microscopy investigation on endomyocardial biopsies. Eur Heart J 2006;27:1847–54.

[38] Beffagna G, Occhi G, Nava A, Vitiello L, Ditadi A, Basso C, Bauce B, Carraro G, Thiene G, Towbin JA, Danieli GA, Rampazzo A. Regulatory mutations in transforming growth factor-beta3 gene cause arrhythmogenic right ventricular cardiomyopathy type 1. Cardiovasc Res 2005;65:366–73.

[39] Merner ND, Hodgkinson KA, Haywood AF, Connors S, French VM, Drenckhahn JD, Kupprion C, Ramadanova K, Thierfelder L, McKenna W, Gallagher B, Morris-Larkin L, Bassett AS, Parfrey PS, Young TL. Arrhythmogenic right ventricular cardiomyopathy type 5 is a fully penetrant, lethal arrhythmic disorder caused by a missense mutation in the TMEM43 gene. Am J Hum Genet 2008;82:809–21.

[40] van Tintelen JP, Van Gelder IC, Asimaki A, Suurmeijer AJ, Wiesfeld AC, Jongbloed JD, van den Wijngaard A, Kuks JB, van Spaendonck-Zwarts KY, Notermans N, Boven L, van den Heuvel F, Veenstra-Knol HE, Saffitz JE, Hofstra RM, van den Berg MP. Severe

cardiac phenotype with right ventricular predominance in a large cohort of patients with a single missense mutation in the DES gene. Heart Rhythm 2009;6:1574–83.

[41] Taylor M, Graw S, Sinagra G, Barnes C, Slavov D, Brun F, Pinamonti B, Salcedo EE, Sauer W, Pyxaras S, Anderson B, Simon B, Bogomolovas J, Labeit S, Granzier H, Mestroni L. Genetic variation in titin in arrhythmogenic right ventricular cardiomyopathy-overlap syndromes. Circulation 2011;124:876–85.

[42] van der Zwaag PA, van Rijsingen IA, Asimaki A, Jongbloed JD, van Veldhuisen DJ, Wiesfeld AC, Cox MG, van Lochem LT, de Boer RA, Hofstra RM, Christiaans I, van Spaendonck-Zwarts KY, Lekanne dit Deprez RH, Judge DP, Calkins H, Suurmeijer AJ, Hauer RN, Saffitz JE, Wilde AA, van den Berg MP, van Tintelen JP. Phospholamban R14del mutation in patients diagnosed with dilated cardiomyopathy or arrhythmogenic right ventricular cardiomyopathy: evidence supporting the concept of arrhythmogenic cardiomyopathy. Eur J Heart Fail 2012;14:1199–207.

[43] Quarta G, Syrris P, Ashworth M, Jenkins S, Zuborne Alapi K, Morgan J, Muir A, Pantazis A, McKenna WJ, Elliott PM. Mutations in the Lamin A/C gene mimic arrhythmogenic right ventricular cardiomyopathy. Eur Heart J 2012;33:1128–36.

[44] van Hengel J, Calore M, Bauce B, Dazzo E, Mazzotti E, De Bortoli M, Lorenzon A, Li Mura IE, Beffagna G, Rigato I, Vleeschouwers M, Tyberghein K, Hulpiau P, van Hamme E, Zaglia T, Corrado D, Basso C, Thiene G, Daliento L, Nava A, van Roy F, Rampazzo A. Mutations in the area composita protein αT-catenin are associated with arrhythmogenic right ventricular cardiomyopathy. Eur Heart J 2013;34:201–10.

[45] Tiso N, Stephan DA, Nava A, Bagattin A, Devaney JM, Stanchi F, Larderet G, Brahmbhatt B, Brown K, Bauce B, Muriago M, Basso C, Thiene G, Danieli GA, Rampazzo A. Identification of mutations in the cardiac ryanodine receptor gene in families affected with arrhythmogenic right ventricular cardiomyopathy type 2 (ARVD2). Hum Mol Genet 2001;10:189–94.

[46] Yang Z, Bowles NE, Scherer SE, Taylor MD, Kearney DL, Ge S, Nadvoretskiy VV, DeFreitas G, Carabello B, Brandon LI, Godsel LM, Green KJ, Saffitz JE, Li H, Danieli GA, Calkins H, Marcus F, Towbin JA. Desmosomal dysfunction due to mutations in desmoplakin causes arrhythmogenic right ventricular dysplasia/cardiomyopathy. Circ Res 2006;99:646–55.

[47] Garcia-Gras E, Lombardi R, Giocondo MJ, Willerson JT, Schneider MD, Khoury DS, Marian AJ. Suppression of canonical Wnt/beta-catenin signaling by nuclear plakoglobin recapitulates phenotype of arrhythmogenic right ventricular cardiomyopathy. J Clin Invest 2006;116:2012–21.

[48] Kirchhof P, Fabritz L, Zwiener M, Witt H, Schäfers M, Zellerhoff S, Paul M, Athai T, Hiller KH, Baba HA, Breithardt G, Ruiz P, Wichter T, Levkau B. Age- and training-dependent development of arrhythmogenic right ventricular cardiomyopathy in heterozygous plakoglobin-deficient mice. Circulation 2006;114:1799–806.

[49] Pilichou K, Remme CA, Basso C, Campian ME, Rizzo S, Barnett P, Scicluna BP, Bauce B, van den Hoff MJ, de Bakker JM, Tan HL, Valente M, Nava A, Wilde AA, Moorman AF, Thiene G, Bezzina CR. Myocyte necrosis underlies progressive myocardial dystrophy in mouse dsg2-related arrhythmogenic right ventricular cardiomyopathy. J Exp Med 2009;206:1802–78.

[50] Rizzo S, Lodder EM, Verkerk AO, Wolswinkel R, Beekman L, Pilichou K, Basso C, Remme CA, Thiene G, Bezzina CR. Intercalated disc abnormalities reduced Na+ current density, and conduction slowing in desmoglein-2 mutant mice prior to cardiomyopathy changes. Cardiovasc Res 2012;95:409–18.

[51] McKenna WJ, Thiene G, Nava A, Fontaliran F, Blomstrom-Lundqvist C, Fontaine G, Camerini F. Diagnosis of arrhythmogenic right ventricular dysplasia/cardiomyopathy. Br Heart J 1994;71:215–8.

[52] Marcus FI, McKenna WJ, Sherrill D, Basso C, Bauce B, Bluemke DA, Calkins H, Corrado D, Cox MG, Daubert JP, Fontaine G, Gear K, Hauer R, Nava A, Picard MH, Protonotarios N, Saffitz JE, Sanborn DM, Steinberg JS, Tandri H, Thiene G, Towbin JA, Tsatsopoulou A,

Wichter T, Zareba W. Diagnosis of arrhythmogenic right ventricular cardiomyopathy/ dysplasia: proposed modification of the task force criteria. Eur Heart J 2010;31:806–14.

[53] Basso C, Ronco F, Marcus F, Abudureheman A, Rizzo S, Frigo AC, Bauce B, Maddalena F, Nava A, Corrado D, Grigoletto F, Thiene G. Quantitative assessment of endomyocardial biopsy in arrhythmogenic right ventricular cardiomyopathy/dysplasia: an in vitro validation of diagnostic criteria. Eur Heart J 2008;29:2760–71.

[54] Marra MP, Leoni L, Bauce B, Corbetti F, Zorzi A, Migliore F, Silvano M, Rigato I, Tona F, Tarantini G, Cacciavillani L, Basso C, Buja G, Thiene G, Iliceto S, Corrado D. Imaging study of ventricular scar in arrhythmogenic right ventricular cardiomyopathy: comparison of 3D standard electroanatomical voltage mapping and contrast-enhanced cardiac magnetic resonance. Circ Arrhythm Electrophysiol 2012;5:91–100.

[55] Corrado D, Basso C, Leoni L, Tokajuk B, Bauce B, Frigo G, Tarantini G, Napodano M, Turrini P, Ramondo A, Daliento L, Nava A, Buja G, Iliceto S, Thiene G. Three-dimensional electroanatomic voltage mapping increases accuracy of diagnosing arrhythmogenic right ventricular cardiomyopathy/dysplasia. Circulation 2005;111:3042–50.

[56] Corrado D, Basso C, Leoni L, Tokajuk B, Turrini P, Bauce B, Migliore F, Pavei A, Tarantini G, Napodano M, Ramondo A, Buja G, Iliceto S, Thiene G. Three-dimensional electroanatomical voltage mapping and histologic evaluation of myocardial substrate in right ventricular outflow tract tachycardia. J Am Coll Cardiol 2008;51:731–9.

[57] Migliore F, Zorzi A, Silvano M, Bevilacqua M, Leoni L, Marra MP, Elmaghawry M, Brugnaro L, Dal Lin C, Bauce B, Rigato I, Tarantini G, Basso C, Buja G, Thiene G, Iliceto S, Corrado D. Prognostic value of endocardial voltage mapping in patients with arrhythmogenic right ventricular cardiomyopathy/dysplasia. Circ Arrhythm Electrophysiol 2013;6:167–76.

[58] Ladyjanskaia GA, Basso C, Hobbelink MG, Kirkels JH, Lahpor JR, Cramer MJ, Thiene G, Hauer RN, V Oosterhout MF. Sarcoid myocarditis with ventricular tachycardia mimicking ARVD/C. J Cardiovasc Electrophysiol 2010;21:94–8.

[59] Corrado D, Basso C, Rizzoli G, Schiavon M, Thiene G. Does sports activity enhance the risk of sudden death in adolescents and young adults? J Am Coll Cardiol 2003;42:1959–63.

[60] Corrado D, Basso C, Pavei A, Michieli P, Schiavon M, Thiene G. Trends in sudden cardiovascular death in young competitive athletes after implementation of a preparticipation screening program. JAMA 2006;296:1593–601.

[61] Migliore F, Zorzi A, Michieli P, Perazzolo Marra M, Siciliano M, Rigato I, Bauce B, Basso C, Toazza D, Schiavon M, Iliceto S, Thiene G, Corrado D. Prevalence of cardiomyopathy in Italian asymptomatic children with electrocardiographic T-wave inversion at preparticipation screening. Circulation 2012;125:529–38.

[62] Thiene G, Rigato I, Pilichou K, Corrado D, Basso C. Arrhythmogenic right ventricular cardiomyopathy. What is needed for a cure? Herz 2012;37:657–62.

[63] Rigato I, Corrado D, Basso C, Zorzi A, Pilichou K, Bauce B, Thiene G. Pharmacotherapy and other therapeutic modalities for managing arrhythmogenic right ventricular cardiomyopathy. Cardiovasc Drugs Ther 2015;29:171–7.

[64] Philips B, Madhavan S, James C, Tichnell C, Murray B, Dalal D, Bhonsale A, Nazarian S, Judge DP, Russell SD, Abraham T, Calkins H, Tandri H. Outcomes of catheter ablation of ventricular tachycardia in arrhythmogenic right ventricular dysplasia/cardiomyopathy. Circ Arrhythm Electrophysiol 2012;5:499–505.

[65] Corrado D, Leoni L, Link MS, Della Bella P, Gaita F, Curnis A, Salerno JU, Igidbashian D, Raviele A, Disertori M, Zanotto G, Verlato R, Vergara G, Delise P, Turrini P, Basso C, Naccarella F, Maddalena F, Estes NA III, Buja G, Thiene G. Implantable cardioverter-defibrillator therapy for prevention of sudden death in patients with arrhythmogenic right ventricular cardiomyopathy/dysplasia. Circulation 2003;108:3084–91.

[66] Corrado D, Calkins H, Link MS, Leoni L, Favale S, Bevilacqua M, Basso C, Ward D, Boriani G, Ricci R, Piccini JP, Dalal D, Santini M, Buja G, Iliceto S, Estes M III, Wichter T, McKenna WJ, Thiene G, Marcus FI. Prophylactic implantable defibrillator in patients

with arrhythmogenic right ventricular cardiomyopathy/dysplasia and no prior ventricular fibrillation or sustained ventricular tachycardia. Circulation 2010;122:1144–52.

[67] Takahashi K, Yamanaka S. Induction of pluripotent stem cells from mouse embryonic and adult fibroblast cultures by defined factors. Cell 2006;126:663–76.

[68] Kim C, Wong J, Wen J, Wang S, Wang C, Spiering C, Kan NG, Forcales S, Puri PL, Leone TC, Marine JE, Calkins H, Kelly DP, Judge DP. Studying arrhythmogenic right ventricular dysplasia with patient-specific iPSCs. Nature 2013;494:105–10.

[69] Murray CE, Pu WT. Reprogramming fibroblasts into cardiomyocytes. New Engl J Med 2011;364:177–8.

[70] Asimaki A, Kapoor S, Plovie E, Karin Arndt A, Adams E, Liu Z, James CA, Judge DP, Calkins H, Churko J, Wu JC, MacRae CA, Kléber AG, Saffitz JE. Identification of a new modulator of the intercalated disc in a zebrafish model of arrhythmogenic cardiomyopathy. Sci Transl Med 2014;6:240. ra74.

[71] Denegri S, Bongianino R, Lodola F, Boncompagni S, De Giusti VC, Avelino-Cruz JE, Liu N, Persampieri S, Curcio A, Esposito F, Pietrangelo L, Marty I, Villani L, Moyaho A, Baiardi P, Auricchio A, Protasi F, Napolitano C, Priori SG. Single delivery of an adeno-associated viral construct to transfer the CASQ2 gene to knock-in mice affected by catecholaminergic polymorphic ventricular tachycardia is able to cure the disease from birth to advanced age. Circulation 2014;129:2673–81.

[72] Bushby K, Lochmuller H, Lynn S, Straub V. Interventions for muscular dystrophy: molecular medicines entering the clinic. Lancet 2009;374:1849–56.

[73] Basso C, Thiene G, Valente M, Angelini A, Corrado D, Nava A. Arrhythmogenic right ventricular cardiomyopathy: dysplasia, dystrophy or myocarditis? Circulation 1996;94:983–91.

[74] Sen-Chowdhry S, Syrris P, Ward D, Asimaki A, Sevdalis E, McKenna WJ. Clinical and genetic characterization of families with arrhythmogenic right ventricular dysplasia/cardiomyopathy provides novel insights into patterns of disease expression. Circulation 2007;115:1710–20.

[75] Basso C, Thiene G. Adipositas cordis, fatty infiltration of the right ventricle, and arrhythmogenic right ventricular cardiomyopathy. Just a matter of fat? Cardiovasc Pathol 2005;14:37–41.

[76] Thiene G, Corrado D, Nava A, Rossi L, Poletti A, Boffa GM, Daliento L, Pennelli N. Right ventricular cardiomyopathy: is there evidence of an inflammatory aetiology? Eur Heart J 1991;12:22–5.

[77] Calabrese F, Basso C, Carturan E, Valente M, Thiene G. Arrhythmogenic right ventricular cardiomyopathy/dysplasia: is there a role for viruses? Cardiovasc Pathol 2006;15:11–7.

[78] Valente M, Calabrese F, Angelini A, Basso C, Thiene G. In vivo evidence of apoptosis in arrhythmogenic right ventricular cardiomyopathy. Am J Pathol 1998;152:479–84.

[79] Pilichou K, Mancini M, Rigato I, Lazzarini E, Giorgi B, Carturan E, Bauce B, d'Amati G, Marra MP, Basso C. Nonischemic left ventricular scar: sporadic or familial? Screen the genes, scan the mutation carriers. Circulation 2014;130:e180–2.

[80] Marcus F, Basso C, Gear K, Sorrell VL. Pitfalls in the diagnosis of arrhythmogenic right ventricular cardiomyopathy/dysplasia. Am J Cardiol 2010;105:1036–9.

[81] Tabib A, Loire R, Chalabreysse L, Meyronnet D, Miras A, Malicier D, Thivolet F, Chevalier P, Bouvagnet P. Circumstances of death and gross and microscopic observations in a series of 200 cases of sudden death associated with arrhythmogenic right ventricular cardiomyopathy and/or dysplasia. Circulation 2003;108:3000–5.

[82] Corrado D, Basso C, Thiene G, McKenna WJ, Davies MJ, Fontaliran F, Nava A, Silvestri F, Blomstrom-Lundqvist C, Wlodarska EK, Fontaine G, Camerini F. Spectrum of clinicopathologic manifestations of arrhythmogenic right ventricular cardiomyopathy/dysplasia: a multicenter study. J Am Coll Cardiol 1997;30:1512–20.

[83] d'Amati G, Leone O, di Gioia CR, Magelli C, Arpesella G, Grillo P, Marino B, Fiore F, Gallo P. Arrhythmogenic right ventricular cardiomyopathy: clinicopathologic correlation based on a revised definition of pathologic patterns. Hum Pathol 2001;32:1078–86.

[84] Aguilera B, Suárez Mier MP, Morentin B. Arrhythmogenic cardiomyopathy as cause of sudden death in Spain. Report of 21 cases. Rev Esp Cardiol 1999;52:656–62.

[85] Angelini A, Basso C, Nava A, Thiene G. Endomyocardial biopsy in arrhythmogenic right ventricular cardiomyopathy. Am Heart J 1996;132:203–6.

[86] Vasaiwala SC, Finn C, Delpriore J, Leya F, Gagermeier J, Akar JG, Santucci P, Dajani K, Bova D, Picken MM, Basso C, Marcus F, Wilber DJ. Prospective study of cardiac sarcoid mimicking arrhythmogenic right ventricular dysplasia. J Cardiovasc Electrophysiol 2009;20:473–6.

[87] Asimaki A, Tandri H, Huang H, Halushka MK, Gautam S, Basso C, Thiene G, Tsatsopoulou A, Protonotarios N, McKenna WJ, Calkins H, Saffitz JE. A new diagnostic test for arrhythmogenic right ventricular cardiomyopathy. N Engl J Med 2009;360: 1075–84.

[88] Asimaki A, Tandri H, Duffy ER, Winterfield JR, Mackey-Bojack S, Picken MM, Cooper LT, Wilber DJ, Marcus FI, Basso C, Thiene G, Tsatsopoulou A, Protonotarios N, Stevenson WG, McKenna WJ, Gautam S, Remick DG, Calkins H, Saffitz JE. Altered desmosomal proteins in granulomatous myocarditis and potential pathogenic links to arrhythmogenic right ventricular cardiomyopathy. Circ Arrhythm Electrophysiol 2011;4:743–52.

3

Definition, Clinical Features, and Classification of ARVC/D. Task Force Criteria for ARVC/D

Aiden Abidov, Arun Kannan, Frank I. Marcus

Department of Medicine/Division of Cardiology
and Department of Medical Imaging,
University of Arizona, Tucson, AZ, USA

DEFINITION OF THE DISEASE

In their 1982 article, Frank Marcus from the University of Arizona and Guy Fontaine with the group of investigators from the *Hôpital de La Salpêtrière* provided a detailed description of the disease they called "arrhythmogenic right ventricular dysplasia," or ARVD [1]. They defined ARVD as *a disorder of heart muscle, leading to arrhythmias, and predominantly involving the right ventricular (RV) myocardium, causing progressive fibro fatty replacement of the RV free wall subsequent RV enlargement and focal regional wall motion abnormalities* (Fig. 3.1). In the original 1982 paper, Marcus et al. described the following clinical picture of the disease: "The typical patient is a *young or middle-aged male* who presents with *palpitations, tachycardia or syncope.* Ventricular tachycardia is of *left bundle branch block configuration.* Physical examination is surprisingly normal, but a fourth heart sound may be present. During sinus rhythm, the ECG may show incomplete or complete right bundle branch block. *T wave inversion in V1 to V4 is usually present. The heart is moderately enlarged and the pulmonary vasculature is normal.*" This description is still applicable to the typical appearance of the main clinical form of this disorder. The authors described in detail imaging-based features of this disorder. Utilizing RV angiography and

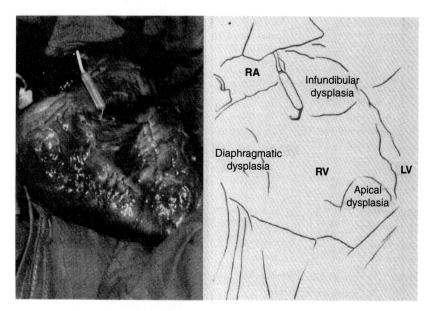

FIGURE 3.1   **The heart of the patient with advanced ARVD during the cardiac surgery.**
Reproduced with permission from Ref. [1].

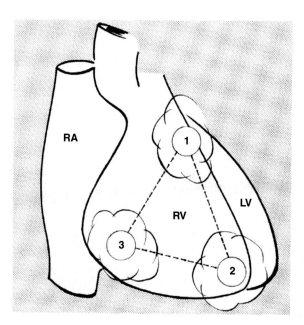

FIGURE 3.2   **Description of the "triangle of dysplasia" provided by Marcus et al.** In the
paper, the authors provided the following legend for the figure: "The most frequent sites of
dysplasia: (1) the anterior infundibulum, (2) the right ventricular apex and (3) the inferior
or diaphragmatic aspect of the right ventricle (RV). These constitute the 'triangle of dysplasia'. LV, left ventricle; RA, right atrium; RV, right ventricle". *Reproduced with permission from
Ref. [1].*

echocardiography, they revealed involvement of specific regions of the RV (RV inflow tract and outflow tract and RV apex) defining it as "the triangle of dysplasia" (Fig. 3.2).

In several recent publications, Dr. Fontaine discussed the controversy associated with the terminology and definition of the disorder, specifically use of the terms "dysplasia" versus "cardiomyopathy" [2,3]. In the original series of the first 24 patients, during surgical ventriculotomy targeting arrhythmias, the investigators described extensive fibrofatty changes of the RV and thinning of the RV free wall. Accordingly, this process was deemed to be a developmental defect (dystrophy or dysplasia). Of interest, Dr. Fontaine noted an abnormal enlarged and thinned RV on one of the anatomic drawings of Leonardo da Vinci (Fig. 3.3). The dysplastic process likely originates *in utero* and gradually develops during adolescence and childhood [3] (Fig. 3.4). However, the clinical spectrum of the arrhythmogenic RV pathology is not limited by a single "typical" dysplastic structural phenotype, and thus, the term "arrhythmogenic right ventricular cardiomyopathy" or ARVC could combine all the disorders with RV anatomic and functional abnormalities and arrhythmogenic potential. Importantly, all these clinical types of ARVC/D follow some "general rules," including slow progression of the disease (Fig. 3.5). Several forms of the pathology, including Carvajal syndrome and Naxos disease are subtypes of cardiocutaneous syndromes [4–6]. Clinical features in these patients consist of a triad of RV cardiomyopathy, woolly hair, and plantar keratoderma (Fig. 3.6). Both clinical disorders are autosomal recessive; while patients with Naxos disease have predominantly RV involvement and nonepidermolytic keratoderma, patients with Carvajal syndrome more frequently have biventricular involvement and epidermolytic cutaneous lesions [6].

Another distinctive form of an RV cardiomyopathy is Uhl's disease, described predominantly in the pediatric population. The original description was an autopsy report in an 8-month-old girl. This anomaly is characterized by the absence of myocardium in the RV free wall that gives the appearance of a "paper-thin" RV wall, or "parchment heart."

The standard definition of ARVC/D usually includes information regarding the autosomal dominant disease inheritance. Genetics of ARVC/D represent an area of active research. These aspects will be discussed in detail in Chapter 6, but from the perspective of the disease definition it is important to understand specific aspects of the disease inheritance. For the typical ARVC/D patient, there is incomplete penetrance and highly variable expressivity [8–10]. Typically, ARVC/D is associated with desmosomal gene mutations, and a total of five desmosomal genes may be found in these patients (plakophilin-2, desmoglein-2, desmocollin, desmoplakin, and junctional plakoglobin) [10]. However, nondesmosomal gene mutations (phospholamban) are not uncommon in patients with ARVC/D and are more frequently associated with biventricular involvement (Fig. 3.7), and with specific ECG findings (of low voltage in the standard leads, and T wave inversion in V3–5) [11].

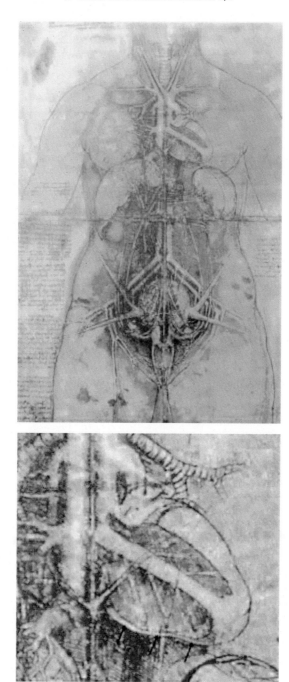

FIGURE 3.3 **Abnormally enlarged and thinned right ventricle on one of the famous drawings by Leonardo Da Vinci noted by Dr. Guy Fontaine.** *Reproduced with permission from Ref. [2].*

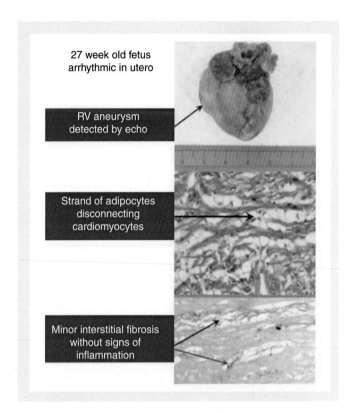

FIGURE 3.4 **Embryology of right ventricular cardiomyopathies.** *Reproduced with permission from Ref. [3].*

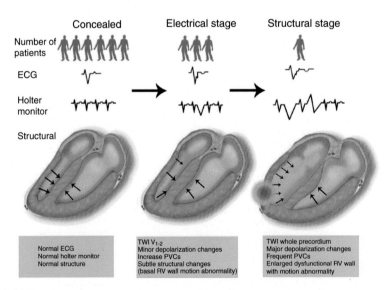

FIGURE 3.5 **Disease progression in the ARVD/C.** ECG, electrocardiogram; PVC, premature ventricular complex; RV, right ventricular; TWI, T-wave inversion. *Reproduced with permission from Ref. [7].*

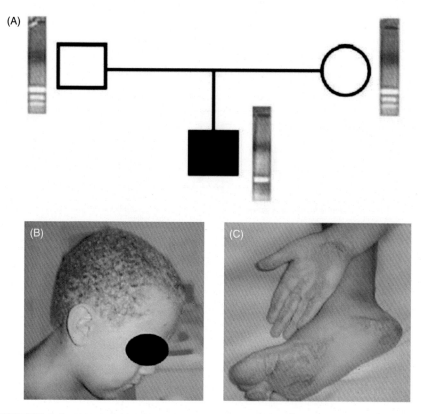

FIGURE 3.6 **Naxos disease: genetic and clinical features.** The original article describes the figure as: (A) restriction analysis revealing the presence of mutant allele (one bold band) in the homozygous boy and the wild-type allele (two light bands) coexisting with the mutant one in the heterozygous carriers parents. (B) woolly hair and (C) keratoderma striate in diffuse plantar areas in the homozygous boy. *Reproduced with permission from Ref. [6].*

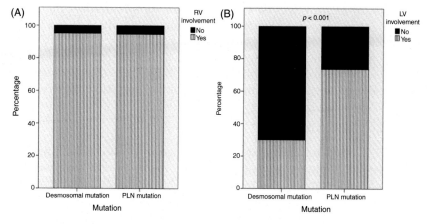

FIGURE 3.7 (A) RV and (B) LV involvement in patients with ARVC/D and desmosomal versus nondesmosomal mutations. PLN, Phospholamban. *Reproduced with permission from Ref. [11].*

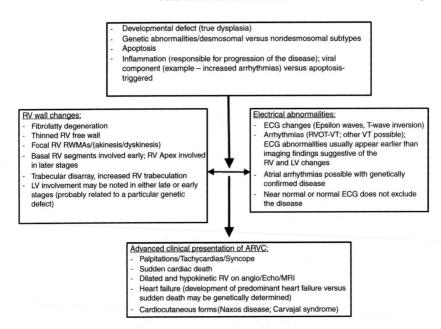

FIGURE 3.8 **Etiopathological considerations in the arrhythmogenic right ventricular cardiomyopathies.** ECG, electrocardiogram; LV, left ventricle; RV, right ventricle; RWMA, regional wall motion abnormality; VT, ventricular tachycardia.

The complexity of the clinical findings as well as diagnostic dilemmas in patients with ARVC/D are presented in Fig. 3.8. We confirm that ARVC/D may be difficult to diagnosis. In this book, we accepted the term "ARVC/D" used in the 2010 Task Force recommendations in order to avoid confusion. As one can see from the earlier discussion, the definition of ARVC/D is not easy. In addition to the autosomal dominant ARVC/D subgroups, there are recessive forms such as Naxos disease. In addition to desmosomal mutations, there are some nondesmosomal gene mutations with different clinical presentations. The disease may be present in either ventricle or have a biventricular involvement. Even the genetics of the disease are complex. However, utilization of the Task Force criteria (TFC) [12] facilitates standardization of the disease and selection of patients with a clinical suspicion for ARVC/D.

Clinical Manifestations

ARVC/D may be identified in adolescence and young adulthood. The diagnosis in a younger population is not common due to the absence of symptoms and signs. The symptoms may have varied presentation and course, differ within the same family [13] and in many instances are progressive [14]. The clinical manifestations of classic ARVC/D include palpitations, lightheadedness, chest pain, dyspnea, syncope, and sudden cardiac death (SCD) [15]. In a prospective study of 108 newly diagnosed patients with suspected ARVC/D, Marcus et al. noted that palpitations was the most common symptom in 56%, followed by dizziness in 27% and syncope in 21% of enrolled patients [15]. Similar observations were noted in a prior study [16]. Due to increased awareness of the disease, asymptomatic younger adults present with a history of ventricular arrhythmias prompting further evaluation. Because of its clinically silent presentation, diagnosing the disease can be difficult, especially in sporadic cases and in the patients with no familial history.

In patients with ARVC/D, palpitations and syncope are the predominant symptoms and are caused by ventricular arrhythmias. The arrhythmias vary from ventricular premature beats to sustained ventricular tachycardia (VT) [17]. Marcus et al. [1] noted that in a group of 22 adults with RV dysplasia, all but one patient had VT with left bundle branch block (LBBB) configuration. In a study that reported the clinical characteristics of 69 living ARVC/D patients, 51 patients had LBBB-type VT and 35 of 52 (69%) had frequent ventricular extrasystoles [17].

The histopathological changes suggestive of ARVC/D may be present in the LV, however, conduction abnormalities are uncommon in ARVC/D [18]. The frequency of ventricular arrhythmias varies depending on the genotype and the severity of disease. In autosomal recessive Naxos disease, ventricular arrhythmias were seen in 92% of adults with the homozygous disease [19]. The incidence of ventricular arrhythmias increases with the extent of cardiac involvement [20].

Supraventricular arrhythmias were seen in 14% patients with ARVC referred for evaluation of ventricular arrhythmias. The SVTs include atrial fibrillation, atrial tachycardia, and atrial flutter [21].

SCD due to ventricular fibrillation may occur in patients with ARVC/D and can be the first clinical presentation of the disease. In a prospective 17 year study of 209 sudden cardiac deaths in athletes and nonathletes (35 years of age or less), in Italy ARVC/D was the most common cause of death (22%) among 49 competitive athletes [22]. However, this relatively high incidence of SCD due to ARVC/D in competitive athletes in Italy has not been observed in reports from other countries [23].

The disease has three distinct pathophysiological phases [7] (Fig. 3.5) that include an early subclinical phase in which no discrete clinical and diagnostic findings are present (concealed). These individuals include children below the ages of 10–12. In the next phase, electrical myocardial changes can be observed without any clinical manifestation of RV dysfunction but patients may have symptomatic ventricular arrhythmias including SCD. The last phase consists of progressively fibrofatty infiltration and replacement of the myocardium leading to RV dilation, dysfunction, aneurysm formation, and right-sided heart failure.

The electrical manifestations are due to histopathological changes. In the early phase, slow conduction abnormalities and electrical uncoupling may occur and this may cause SCD. As the disease advances there is heterogeneous fibrofatty infiltration, resulting in increased arrhythmogenicity. As mentioned earlier, the predominant area of RV involvement includes "the triangle of dysplasia" that results in inverted T waves in the anterior precordial leads to a characteristic monomorphic VT with LBBB morphology.

In patients with arrhythmogenic left ventricular cardiomyopathy, symptoms are similar to right-sided ARVC/D, but syncope is less common [24]. The extracardiac manifestations of ARVC/D may include woolly hair, palmar, and plantar keratoderma and are seen in Naxos disease [25].

Due to the atypical presentation and nonspecific symptoms, diagnosis of ARVC/D is often challenging. The myocardial biopsy is often considered the gold standard in diagnosing ARVC/D but it is also marred by limitations [26]. ARVC/D typically involves the RV free wall and the risk of perforation is small but higher when the myocardial biopsy is attempted in the free wall and so is usually done in the IV septum decreasing the yield. Also the biopsy is invasive and the fibrofatty infiltrations seen in ARVC/D can also be seen in other conditions [27]. Finally, ARVC/D is focal and the biopsy may miss the affected area of the RV.

In general, ARVC/D should be suspected when young or middle aged patients present with symptomatic or asymptomatic VT with LBBB morphology without clinical evidence of coronary artery disease and any other structural heart disease, in survivors of SCDs, and in patients with incidental finding of an RV abnormality by echocardiograms or MRI.

Task Force Criteria

Due to lack of specific symptoms and signs, TFC for the clinical diagnosis of ARVC/D was proposed in 1994 [28]. There was agreement that there is no single diagnostic modality to confirm the disease. The multitude of criteria included structural, histological, ECG, arrhythmic, and familial features of the disease. The abnormalities were categorized into major and minor criteria. Conventional signs suggestive of ARVC/D including complete and incomplete RBBB, and arrhythmias originating from the RV were not included in the criteria due to the lack of specificity. The echocardiographic findings of RV were subjective and nonspecific. Also some of the RV changes such as wall motion abnormalities were frequently seen in healthy subjects [29].

Due to the discovery of the genetics involved in ARVC/D and advances in cardiac imaging techniques such as MRI, modifications of the criteria have been proposed to improve the diagnosis of ARVC/D in family members and in people with incomplete expression of the disease [12,30]. The revised 2010 criteria [12] contain the same categories proposed by the original TFC but include more specific assessment in each category, thus improving the specificity and sensitivity of the criteria (Table 3.1).

The revised TFC for the diagnosis of ARVC/D are comprehensive and include objective morphological and clinical characteristics of ARVC/D. The modifications were also aimed to facilitate the diagnosis in first-degree relatives with incomplete expression and to quantify RV dimensions [31]. The structural changes in the RV were modified to include RVOT dimensions and fractional area change by echocardiogram and RV end-diastolic volume and RV ejection fraction by MRI. Also the cardiac abnormalities were made more specific with morphometric analysis compared to older criteria, which included the confirmation of fibrofatty replacement based on endomyocardial biopsy. Likewise there are more specific criteria for repolarization, depolarization abnormalities, and arrhythmias. Due to the discoveries of the genetics of ARVC/D and identification of the familial members, significant changes were made for the identification of the disease in first-degree relatives.

The diagnosis of ARVC/D is fulfilled by the presence of two major, or one major plus two minor criteria or four minor criteria from different groups [12]. Diagnostic terminology based on TFC is summarized as follows:

1. Definitive diagnosis: two major or one major and two minor criteria or four minor from different categories;
2. Borderline diagnosis: one major and one minor or three minor criteria from different categories;
3. Possible diagnosis: one major or two minor criteria from different categories.

Establishing the diagnosis of ARVC/D remains a clinical challenge, as there is no single test to establish the identity of the disease due to its varied genotypic and phenotypic presentation. It should be noted that the cutoff values for RV volumetrics are different according to age [35], raising the concerns of universality of the criteria in diagnosing ARVC/D. The current modified TFC remains the most comprehensive method to diagnose ARVC/D.

ARVC/D was included as a separate term in the 2006 AHA genetics-based classification of the cardiomyopathies as well as in the 2008 European Society of Cardiology classification, with phenotype and genotype considerations (Fig. 3.9). Most recently, the ARVC/D,

TABLE 3.1   Comparison of the Original and Revised Task Force ARVC/D Criteria

Original TFC	Revised TFC

1. Global or regional dysfunction and structural alterations*

Major

• Severe dilatation and reduction of RV ejection fraction with no (or only mild) LV impairment
• Localized RV aneurysms (akinetic or dyskinetic areas with diastolic bulging)
• Severe segmental dilatation of the RV
	By 2D echo
	• Regional RV akinesia, dyskinesia, or aneurysm
	• *and* one of the following (end diastole):
	  • PLAX RVOT ≥32 mm (corrected for body size [PLAX/BSA] ≥19 mm/m^2)
	  • PSAX RVOT ≥36 mm (corrected for body size [PSAX/BSA] ≥21 mm/m^2)
	  • *or* fractional area change ≤33%

	By MRI
	• Regional RV akinesia or dyskinesia or dyssynchronous RV contraction
	• *and* one of the following:
	  — Ratio of RV end-diastolic volume to BSA ≥110 mL/m^2 (male) or ≥100 mL/m^2 (female)
	  — *or* RV ejection fraction ≤40%

	By RV angiography • Regional RV akinesia, dyskinesia, or aneurysm

Minor

• Mild global RV dilatation and/or ejection fraction reduction with normal LV
• Mild segmental dilatation of the RV
• Regional RV hypokinesia
	By 2D echo
	• Regional RV akinesia or dyskinesia
	• *and* one of the following (end diastole):
	  • PLAX RVOT ≥29 to <32 mm (corrected for body size [PLAX/BSA] ≥16 to <19 mm/m^2)
	  • PSAX RVOT ≥32 to <36 mm (corrected for body size [PSAX/BSA] ≥18 to <21 mm/m^2)
	  • *or* fractional area change >33% to ≤40%

	By MRI
	• Regional RV akinesia or dyskinesia or dyssynchronous RV contraction
	• *and* one of the following:
	  — Ratio of RV end-diastolic volume to BSA ≥100 to <110 mL/m^2 (male) or ≥90 to <100 mL/m^2 (female)
	  — *or* RV ejection fraction >40% to ≤45%

2. Tissue characterization of wall

Major

• Fibrofatty replacement of myocardium on endomyocardial biopsy	• Residual myocytes <60% by morphometric analysis (or <50% if estimated), with fibrous replacement of the RV free wall myocardium in ≥1 sample, with or without fatty replacement of tissue on endomyocardial biopsy

(Continued)

TABLE 3.1   Comparison of the Original and Revised Task Force ARVC/D Criteria (*cont.*)

Original TFC	Revised TFC
Minor	
	• Residual myocytes 60% to 75% by morphometric analysis (or <50% to <65% if estimated), with fibrous replacement of the RV free wall myocardium in ≥1 sample, with or without fatty replacement of tissue on endomyocardial biopsy

3. Repolarization abnormalities

Major	
	• Inverted T waves in right precordial leads (V_1, V_2, and V_3) or beyond in individuals >14 years of age (in the absence of complete RBBB QRS ≥120 ms)
Minor	
• Inverted T waves in right precordial leads (V_2 and V_3) (people age >12 years, in absence of RBBB)	• Inverted T waves in leads V_1 and V_2 in individuals >14 years of age (in the absence of complete RBBB) or in V_4, V_5, or V_6 • Inverted T waves in leads V_1, V_2, V_3, and V_4 in individuals >14 years of age in the presence of complete RBBB

4. Depolarization/conduction abnormalities

Major	
• Epsilon waves or localized prolongation (>110 ms) of the QRS complex in right precordial leads (V_1–V_3)	• Epsilon wave (reproducible low-amplitude signals between end of QRS complex to onset of the T wave) in the right precordial leads (V_1–V_3)
Minor	
• Late potentials (SAECG)	• Late potentials by SAECG in ≥1 of 3 parameters in the absence of a QRS duration of ≥110 ms on the standard ECG • Filtered QRS duration (fQRS) ≥114 ms • Duration of terminal QRS <40 μV (low-amplitude signal duration) ≥38 ms • Root-mean-square voltage of terminal 40 ms ≤20 μV • Terminal activation duration of QRS ≥55 ms measured from the nadir of the S wave to the end of the QRS, including R′, in V_1, V_2, or V_3, in the absence of complete RBBB

5. Arrhythmias

Major	
	• Nonsustained or sustained VT of left bundle-branch morphology with superior axis (negative or indeterminate QRS in leads II, III, and aVF and positive in lead aVL)

TABLE 3.1 Comparison of the Original and Revised Task Force ARVC/D Criteria *(cont.)*

Original TFC	Revised TFC
Minor	
• Left bundle-branch block-type VT (sustained and nonsustained) (ECG, Holter, exercise) • Frequent ventricular extrasystoles (>1000 per 24 h) (Holter)	• Nonsustained or sustained VT of RV outflow configuration, left bundle-branch block morphology with inferior axis (positive QRS in leads II, III, and aVF and negative in lead aVL) or of unknown axis • >500 ventricular extrasystoles per 24 h (Holter)
6. Family history	
Major	
• Familial disease confirmed at necropsy or surgery	• ARVC/D confirmed in a first-degree relative who meets current TFC criteria • ARVC/D confirmed pathologically at autopsy or surgery in a first-degree relative • Identification of a pathogenic mutation** categorized as associated or probably associated with ARVC/D in the patient under evaluation
Minor	
• Family history of premature sudden death (<35 years of age) due to suspected ARVC/D • Familial history (clinical diagnosis based on present criteria)	• History of ARVC/D in a first-degree relative in whom it is not possible or practical to determine whether the family member meets current TFC criteria • Premature sudden death (<35 years of age) due to suspected ARVC/D in a first-degree relative • ARVC/D confirmed pathologically or by current TFC criteria in second-degree relative

PLAX indicates parasternal long-axis view; RVOT, RV outflow tract; BSA, body surface area; PSAX, parasternal short-axis view; aVF, augmented voltage unipolar left foot lead; aVL, augmented voltage unipolar left arm lead.

Diagnostic terminology for original criteria: This diagnosis is fulfilled by the presence of two major, or one major plus two minor criteria, or four minor criteria from different groups. Diagnostic terminology for revised criteria: definite diagnosis: two major or one major and two minor criteria or four minor from different categories; borderline: one major and one minor or three minor criteria from different categories; possible: one major or two minor criteria from different categories.

Hypokinesis is not included in this or subsequent definitions of RV regional wall motion abnormalities for the proposed modified criteria.

**A pathogenic mutation is a DNA alteration associated with ARVC/D that alters or is expected to alter the encoded protein, is unobserved or rare in a large non-ARVC/D control population, and either alters or is predicted to alter the structure or function of the protein or has demonstrated linkage to the disease phenotype in a conclusive pedigree. Reproduced, with permission, from Ref. [12].*

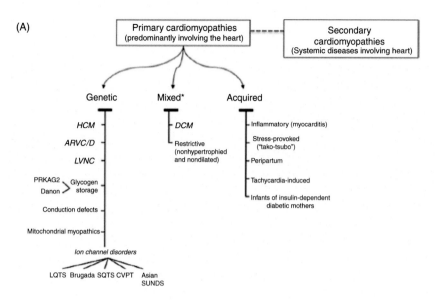

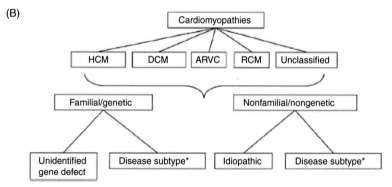

FIGURE 3.9 **Current clinical classification of cardiomyopathies.** (A) The 2006 American Heart Association classification utilized genetics-based classification of cardiomyopathies; (B) The 2008 European Society of Cardiology classification utilized the morphofunctional phenotype and genotype information. ARVC/D, Arrhythmogenic right ventricular cardiomyopathy/dysplasia; CVPT, catecholaminergic polymorphic ventricular tachycardia; DCM, dilated cardiomyopathy; HCM, hypertrophic cardiomyopathy; LVNC, left ventricular noncompaction; LQTS, long QT syndrome; RCM, restrictive cardiomyopathy; SQTS, short QT syndrome; SUNDS, sudden unexplained nocturnal death syndrome. *Reproduced with permission from Ref. [36].*

among other cardiomyopathies, was also included in the MOGE(S) – morphofunctional phenotype (M); organ (O); genetics (G); etiology (E); and clinical stage (S) (see Table 3.2) [36]. The classification allows a great degree of standardization in reporting diagnostic information; however, it may be considered too complex for implementation in daily clinical practice.

TABLE 3.2 The MOGE(S) Classification of Cardiomyopathies

M Morphofunctional phenotype*	O Organ/system involvement**	G Genetic†	E Etiological annotation‡	S Stage; ACC/AHA Stage, NYHA functional class§
(D) Dilated (H) Hypertrophic (R) Restrictive (A) ARVC (NC) LVNC Overlapping (H + R), (D + A), (NC + H), (H + D), (D + NC) or more complex combinations such as (H + R + NC) (E) Early, with type in parentheses (NS) Nonspecific phenotype (NA) Information not available (O) Unaffected	(H) Heart (M) Muscle, skeletal (N) Nervous (C) Cutaneous (E) Eye (A) Auditory (K) Kidney (G) Gastrointestinal (S) Skeletal (O) Absence of organ/system involvement, for example, in family members who are healthy mutation carriers; the mutation is specified in E and inheritance in G	(N) Family history negative (U) Family history unknown (AD) Autosomal dominant (AR) Autosomal recessive (XLR) X-linked recessive (XLD) X-linked dominant (XL) X-linked (M) Matrilineal (DN) De novo (O) Family history not investigated	(G) Genetic etiology – add gene and mutation; (NC) Individual noncarrier plus the gene that tested negative (OC) Obligate carrier (ONC) Obligate noncarrier (DN) De novo (C) Complex genetics when >1 mutation (provide additional gene and mutation) (Neg) Genetic test negative for the known familial mutation (NA) Genetic test not yet available (N) Genetic defect not identified (O) No genetic test, any reason (no blood sample, no informed consent, etc.) Genetic amyloidosis (A-TTR) or hemochromatosis (HFE) Nongenetic etiologies: (M) Myocarditis (V) Viral infection (add the virus identified in affected heart) (AI) Autoimmune/immune-mediated; suspected (AI-S), proven (AI-P); (A) Amyloidosis (add type of amyloidosis: A-K; A-L, H-SAA); (I) Infectious, nonviral (add the infectious agent) (T) Toxicity (add toxic cause/drug) (Eo) Hypereosinophilic heart disease	ACC/AHA stage represented as letter (A, B, C, D) To be followed by NYHA functional class represented in Roman numerals (I, II, III, IV)
M_D M_H M_R M_A M_{NC}, M_O M_{H+R} M_{D+A}	O_H, O_M O_K, O_C	G_N, G_U G_{AD}, G_{AR}, G_{XLR} G_{XLD}, G_{XD}, G_M G_{DN}	$E_{G-MYH7[R403E]}$, $E_{G-HFE[Cys282Tyr+/+]}$ E_{V-HCMV} $E_{G-A-TTR[V30M]}$ $E_{M-sarcoidosis}$	S_{A-I}, S_{A-II}

ACC/AHA, American College of Cardiology – American Heart Association; NYHA, New York Heart Association; other abbreviations as in Table 3.1.

*The morphofunctional phenotype description (M) may contain more information using standard abbreviations, such as AVB = atrioventricular block; WPW = Wolff-Parkinson-White syndrome; LQT = prolongation of the QT interval; AF = atrial fibrillation; ↓R = low electrocardiogram voltages; ↓PR = short PR interval.

**Organ (O) involvement in addition to the H subscript (for heart) should be expanded (for heart) for the involvement of M = skeletal muscle, E = eye, ocular system, A = auditory system, K = kidney, L = liver, N = nervous system, C = cutaneous, G = gastrointestinal system, and other comorbidities, including MR = mental retardation.

†Genetic (G) describes the available information about inheritance of the disease. It also provides complete information if the family history is not proven or unknown, and if genetic testing has not been performed or was negative for the mutation/mutations identified in the family.

‡The etiologic annotation (E) provides the facility for the synthetic description of the specific disease gene and mutation, as well as description of nongenetic etiology.

§The functional annotation or staging (S) allows the addition of ACC/AHA stage and NYHA functional class.

Reproduced with permission from Ref. [36].

References

[1] Marcus FI, Fontaine GH, Guiraudon G, et al. Right ventricular dysplasia: a report of 24 adult cases. Circulation 1982;65:384–98.

[2] Fontaine G, Chen HS. Arrhythmogenic right ventricular dysplasia back in force. Am J Cardiol 2014;113:1735–9.

[3] Fontaine GH. The multiple facets of right ventricular cardiomyopathies. Eur Heart J 2011;32:1049–51.

[4] Protonotarios N, Tsatsopoulou A, Fontaine G. Naxos disease: keratoderma, scalp modifications, and cardiomyopathy. J Am Acad Dermatol 2001;44:309–11.

[5] Baykan A, Olgar S, Argun M, et al. Different clinical presentations of Naxos disease and Carvajal syndrome: case series from a single tertiary center and review of the literature. Anatol J Cardiol 2015;15(5):404–8.

[6] Protonotarios N, Tsatsopoulou A. Naxos disease and Carvajal syndrome: cardiocutaneous disorders that highlight the pathogenesis and broaden the spectrum of arrhythmogenic right ventricular cardiomyopathy. Cardiovasc Pathol 2004;13:185–94.

[7] te Riele AS, James CA, Rastegar N, et al. Yield of serial evaluation in at-risk family members of patients with ARVD/C. J Am Coll Cardiol 2014;64:293–301.

[8] te Riele AS, Bhonsale A, James CA, et al. Incremental value of cardiac magnetic resonance imaging in arrhythmic risk stratification of arrhythmogenic right ventricular dysplasia/cardiomyopathy-associated desmosomal mutation carriers. J Am Coll Cardiol 2013;62:1761–9.

[9] Dalal D, James C, Devanagondi R, et al. Penetrance of mutations in plakophilin-2 among families with arrhythmogenic right ventricular dysplasia/cardiomyopathy. J Am Coll Cardiol 2006;48:1416–24.

[10] Tan BY, Jain R, den Haan AD, et al. Shared desmosome gene findings in early and late onset arrhythmogenic right ventricular dysplasia/cardiomyopathy. J Cardiovasc Transl Res 2010;3:663–73.

[11] Groeneweg JA, van der Zwaag PA, Olde Nordkamp LR, et al. Arrhythmogenic right ventricular dysplasia/cardiomyopathy according to revised 2010 task force criteria with inclusion of non-desmosomal phospholamban mutation carriers. Am J Cardiol 2013;112:1197–206.

[12] Marcus FI, McKenna WJ, Sherrill D, et al. Diagnosis of arrhythmogenic right ventricular cardiomyopathy/dysplasia: proposed modification of the task force criteria. Circulation 2010;121:1533–41.

[13] Nava A, Scognamiglio R, Thiene G, et al. A polymorphic form of familial arrhythmogenic right ventricular dysplasia. Am J Cardiol 1987;59:1405–9.

[14] Blomstrom-Lundqvist C, Sabel KG, Olsson SB. A long term follow up of 15 patients with arrhythmogenic right ventricular dysplasia. Br Heart J 1987;58:477–88.

[15] Marcus FI, Zareba W, Calkins H, et al. Arrhythmogenic right ventricular cardiomyopathy/dysplasia clinical presentation and diagnostic evaluation: results from the North American Multidisciplinary Study. Heart Rhythm 2009;6:984–92.

[16] Hulot JS, Jouven X, Empana JP, Frank R, Fontaine G. Natural history and risk stratification of arrhythmogenic right ventricular dysplasia/cardiomyopathy. Circulation 2004;110:1879–84.

[17] Dalal D, Nasir K, Bomma C, et al. Arrhythmogenic right ventricular dysplasia: a United States experience. Circulation 2005;112:3823–32.

[18] Tabib A, Loire R, Chalabreysse L, et al. Circumstances of death and gross and microscopic observations in a series of 200 cases of sudden death associated with arrhythmogenic right ventricular cardiomyopathy and/or dysplasia. Circulation 2003;108:3000–5.

[19] Protonotarios N, Tsatsopoulou A, Anastasakis A, et al. Genotype-phenotype assessment in autosomal recessive arrhythmogenic right ventricular cardiomyopathy (Naxos disease) caused by a deletion in plakoglobin. J Am Coll Cardiol 2001;38:1477–84.

[20] Nava A, Bauce B, Basso C, et al. Clinical profile and long-term follow-up of 37 families with arrhythmogenic right ventricular cardiomyopathy. J Am Coll Cardiol 2000;36: 2226–33.

[21] Camm CF, James CA, Tichnell C, et al. Prevalence of atrial arrhythmias in arrhythmogenic right ventricular dysplasia/cardiomyopathy. Heart Rhythm 2013;10:1661–8.

[22] Corrado D, Basso C, Schiavon M, Thiene G. Screening for hypertrophic cardiomyopathy in young athletes. N Engl J Med 1998;339:364–9.

[23] Hedrich O, Este M III, Link M. Sudden cardiac death in athletes. Curr Cardiol Rep 2006;8:316–22.

[24] Sen-Chowdhry S, Syrris P, Prasad SK, et al. Left-dominant arrhythmogenic cardiomyopathy: an under-recognized clinical entity. J Am Coll Cardiol 2008;52:2175–87.

[25] Coonar AS, Protonotarios N, Tsatsopoulou A, et al. Gene for arrhythmogenic right ventricular cardiomyopathy with diffuse nonepidermolytic palmoplantar keratoderma and woolly hair (Naxos disease) maps to 17q21. Circulation 1998;97:2049–58.

[26] Basso C, Thiene G, Corrado D, Angelini A, Nava A, Valente M. Arrhythmogenic right ventricular cardiomyopathy. Dysplasia, dystrophy, or myocarditis? Circulation 1996;94: 983–91.

[27] Unverferth DV, Baker PB, Swift SE, et al. Extent of myocardial fibrosis and cellular hypertrophy in dilated cardiomyopathy. Am J Cardiol 1986;57:816–20.

[28] McKenna WJ, Thiene G, Nava A, et al. Diagnosis of arrhythmogenic right ventricular dysplasia/cardiomyopathy. Task Force of the Working Group Myocardial and Pericardial Disease of the European Society of Cardiology and of the Scientific Council on Cardiomyopathies of the International Society and Federation of Cardiology. Br Heart J 1994;71:215–8.

[29] Sievers B, Addo M, Franken U, Trappe HJ. Right ventricular wall motion abnormalities found in healthy subjects by cardiovascular magnetic resonance imaging and characterized with a new segmental model. J Cardiovasc Magn Reson 2004;6:601–8.

[30] Hamid MS, Norman M, Quraishi A, et al. Prospective evaluation of relatives for familial arrhythmogenic right ventricular cardiomyopathy/dysplasia reveals a need to broaden diagnostic criteria. J Am Coll Cardiol 2002;40:1445–50.

[31] Bluemke DA. ARVC: imaging diagnosis is still in the eye of the beholder. JACC Cardiovasc Imaging 2011;4:288–91.

[32] Lindsay BD. Challenges of diagnosis and risk stratification in patients with arrhythmogenic right ventricular cardiomyopathy/dysplasia. J Am Coll Cardiol 2013;62:1770–1.

[33] Sen-Chowdhry S, Prasad SK, Syrris P, et al. Cardiovascular magnetic resonance in arrhythmogenic right ventricular cardiomyopathy revisited: comparison with task force criteria and genotype. J Am Coll Cardiol 2006;48:2132–40.

[34] Vermes E, Strohm O, Otmani A, Childs H, Duff H, Friedrich MG. Impact of the revision of arrhythmogenic right ventricular cardiomyopathy/dysplasia task force criteria on its prevalence by CMR criteria. JACC Cardiovasc Imaging 2011;4:282–7.

[35] Chahal H, Johnson C, Tandri H, et al. Relation of cardiovascular risk factors to right ventricular structure and function as determined by magnetic resonance imaging (results from the multi-ethnic study of atherosclerosis). Am J Cardiol 2010;106:110–6.

[36] Arbustini E, Narula N, Dec GW, et al. The MOGE(S) classification for a phenotype-genotype nomenclature of cardiomyopathy: endorsed by the World Heart Federation. Global heart 2013;8:355–82.

CMR Features of ARVC/D

*Aiden Abidov, Arun Kannan, Isabel B. Oliva,
Frank I. Marcus*

Department of Medicine/Division of Cardiology
and Department of Medical Imaging,
University of Arizona, Tucson, AZ, USA

DIAGNOSING ARVC/D BY CMR

Since George Diamond et al. postulated the use of pretest probability for diagnostic testing [1,2], this principle is being utilized in multiple fields of cardiac imaging. Utilization of pretest probability/likelihood of disease may influence the interpretation and diagnostic certainty of the reader. An example of this diagnostic process is the application of the pretest likelihood of coronary arterial disease (CAD) in nuclear cardiology [3] and coronary CTA [4]. Similarly, the diagnosis of the right ventricular (RV) cardiomyopathies may be influenced by the assessment of the pretest probability of ARVC/D, and the cardiac MRI (CMR) protocol that is planned for the patient. When CMR is ordered to confirm the possibility of a high probability of ARVC/D, the physician interpreting the CMR should apply his/her "sensitivity reading" skills, describing any subtle abnormalities noted on the scan. On the other hand, when the pretest likelihood of the ARVC/D is low, CMR may be utilized as the "specificity tool," effectively increasing the overall negative predictive value of this diagnostic modality.

CMR results are often needed as an "ultimate" diagnostic test, since only 50–60% of genetic tests are positive in patients with ARVC/D. Even with this information, one cannot state that the patient has a clinically relevant disorder. Adding to this diagnostic complexity is the potential risk of sudden death associated with ARVC/D. This is especially important, since the majority of patients referred to CMR for suspected ARVC/D are young or middle aged individuals.

The utility of CMR for the diagnosis of ARVC/D has been evaluated in numerous studies and is an integral part of the Task Force criteria (TFC) [5]

TABLE 4.1 Definition of Major and Minor CMR Criteria According to the Original 1994 Task Force Criteria and the 2010 Revised Task Force Criteria

Criteria	Original criteria – 1994	Revised criteria – 2010
Major	Severe RV dilatation and reduced RVEF (normal LV) or localized RV aneurysms or severe segmental RV dilatation	Regional RV akinesia or regional dyskinesia or dyssynchronous RV contraction and RVEDVI/BSA $\geq$ 110 mL/m^2 (male) or RVEDVI/BSA $\geq$ 100 mL/m^2 (female) or RVEF $\leq$ 40%
Minor	Mild global RV dilatation and/or reduced RVEF (normal LV) or regional RV hypokinesia or mild segmental RV dilatation	Regional RV akinesia or regional dyskinesia or dyssynchronous RV contraction and RVEDVI/BSA 100 to 109 mL/m^2 (male) or RVEDVI/BSA 90 to 99 mL/m^2 (female) or RVEF 41% to 45%

BSA, body surface area; LV, left ventricle; RV, right ventricle; RVEDV, right ventricular end-diastolic volume; RVEDVI, right ventricular end-diastolic volume index (indexed to BSA); RVEF, right ventricular ejection fraction.
Reproduced with permission from Refs [10,11].

(Table 4.1). CMR appears to have a role in identifying (Fig. 4.1) subtle early changes of ARVC/D with sensitivity as high as 100% and a specificity of about 50% [6]. It also has incremental diagnostic value in risk stratifying asymptomatic ARVC/D-associated desmosomal mutation carriers [7] (Fig. 4.2). CMR data also have an important prognostic value in ARVC/D [8,9] (Table 4.2). This will be discussed in Chapter 8.

We recommend carefully evaluating the patient's clinical history as well as results of any testing prior to the index CMR. This is one of the mandatory rules for cardiac imagers supervising the scan for patients with known or suspected ARVC/D. When the patient has a history of palpitations and syncope, family history of sudden death, abnormal ECG with inverted T waves in right precordial leads, and evidence of dilated RV on a recent echocardiogram, the level of pretest likelihood of this patient having ARVC/D is very high. Accordingly, the significance of any CMR abnormality in the patient is weighed differently from the patient with a low pretest likelihood of ARVC/D. If the pretest likelihood of ARVC/D is high, but the CMR appears normal, the cardiologist should consider additional testing and close follow-up to make sure that the patient is safe. This may focus on electrical abnormalities such as the ECG or Holter monitoring. It is important to note that the diagnosis of ARVC/D does not require an abnormal CMR.

Various imaging techniques are available in evaluating the morphological changes of ARVC/D and include two-dimensional transthoracic echocardiography, CMR radionuclide angiography, multidetector cardiac CT, and RV angiography. The conventional transthoracic two-dimensional

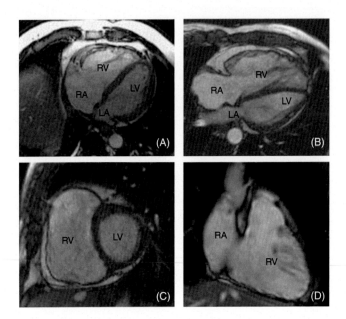

FIGURE 4.1 **ARVC/D appearance on CMR images.** The images represent a balanced steady-state free precession (bSSFP) sequences of the patients with confirmed ARVC/D. On all the images, the RV is dilated; cine images confirmed presence of the RV systolic dysfunction, with RVEF <40%. (A) axial view predominant basal/subtricuspid enlargement of the RV with relatively spared RV apex; (B, C) axial and short-axis views showing diffuse RV enlargement with multiple small focal RV free wall outpouchings; (D) dedicated RV inflow/outflow view, showing multiple small focal RV posterior wall outpouchings and focal RV outflow aneurysm.

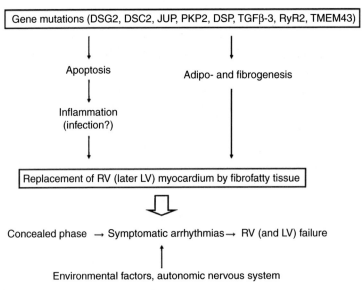

FIGURE 4.2 **Pathogenesis of ARVC/D.** Legend from the original publication: DSC2, desmocollin-2; DSG2, desmoglein-2; DSP, desmoplakin; JUP, plakoglobin; LV, left ventricular; PKP2, plakophilin-2; RV, right ventricular; RyR2, cardiac ryanodine receptor, TGFb-3, transforming growth factor b-3; TMEM43, transmembrane protein 43. *Reproduced with permission from Ref. [12].*

TABLE 4.2   Predictors of Adverse Outcome in Patients With ARVC/D

History
   Malignant familial background (sudden death)
   Previous cardiac arrest
   Syncope or sustained ventricular tachycardia, especially if hemodynamically unstable
     (impairment of consciousness) or pleomorphic
   Early onset of symptoms
   History of congestive heart failure
ECG
   Prolonged PR duration
   Increased QRS dispersion (a difference of ≥40 ms between the maximum and
     minimum QRS durations occurring in any of the 12 ECG leads)
   Increased QT dispersion (≥65 ms)
   Presence of bundle branch block
   Epsilon waves or late potentials
Echocardiography or cardiac magnetic resonance imaging
   Severe right ventricular involvement and dilation
   Left ventricular involvement
   Impaired function of the right (and left) ventricle
   Left atrial dilation

Reproduced with permission from Ref. [12].

echocardiogram (TTE) and three-dimensional TTE are readily available with no radiation exposure. The echocardiogram is useful to assess regional RV akinesia, dyskinesia, or focal aneurysms and assess the RV dimensions that are an important part of the TFC [5]. However, the diagnostic accuracy decreases when there is a poor acoustic window, especially with a dilated RV [13]. Cardiac CTA (CCTA) can also aid in assessing the myocardial changes in ARVC/D such as RV dilatation, prominent trabeculation, and a scalloped surface of the RV free wall by providing three-dimensional assessment [14]. Clinical use of the CCTA, especially in young patients, is limited by substantial radiation exposure. In the past, conventional right heart angiography was considered a standard imaging modality for the diagnosis of ARVC/D and is still useful to assess wall motion abnormalities (WMAs) of the RV but it is invasive with potential vascular complications [15].

CMR offers a unique noninvasive approach to assess the cardiac anatomy and function of the RV as well as analyzing the flow through the pulmonary artery (PA) and the aorta. Due to its excellent spatial and temporal resolution as well as its ability to obtain extracardiac imaging data with cardiac acquisition, CMR is rapidly gaining widespread acceptance to evaluate patients with suspected ARVC/D since it has the capability to assess tissue characteristics such as differentiating muscle from fat [16]. Also CMR has no radiation exposure and can be used in patients with renal dysfunction, except those with end-stage renal disease.

CMR FINDINGS IN ARVC/D: TYPICAL IMAGING FEATURES

The typical CMR findings in ARVC/D include RV dilatation and global RV systolic dysfunction, RV regional dysfunction/regional wall motion abnormalties (WMAs), trabecular disarray, thinned and remodeled RV myocardium due to fibrofatty replacement and aneurysms of the RV and RVOT, and delayed hyperenhancement (DHE) of the involved myocardium on the gadolinium (Gd)-enhanced sequences (Fig. 4.3).

Current ARVC/D TFC require the presence of qualitative WMA and increased indexed EDV or reduced ejection fraction (Fig. 4.4). If either RV size or RVEF is positive and there is RV WMA, the patient would be classified as having a major criterion. Sensitivity of the RV size alone or the RV dysfunction alone is reported to range from 41% to 50% for major criteria and 31% to 41% for minor criteria. The sensitivity increases from 79% to 89% for major criteria and from 68% to 78% for minor criteria when using indexed RVEDV (area) or RVEF (function) [5].

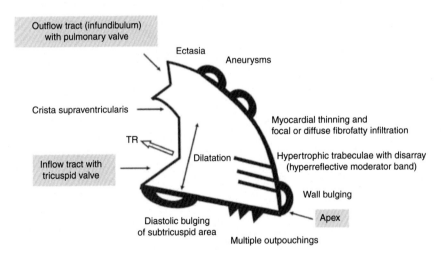

FIGURE 4.3 **Summary of the significant imaging findings in patients with ARVC/D.** *Reproduced with permission from Ref. [12].*

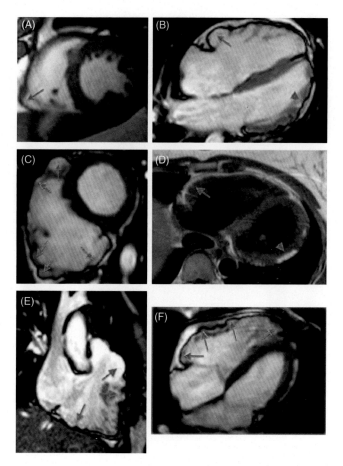

FIGURE 4.4 **CMR evidence of the focal and regional RV systolic dysfunction and abnormal RV structure in ARVC/D.** Legend from the original article [7]: (A) Short-axis bright blood image showing dyskinesis of the acute angle (arrow). (B) Horizontal long-axis bright blood image showing subtricuspid bulging (arrow). Subepicardial fatty infiltration and wall thinning of the LV apicolateral region (arrowhead) is present. (C) SAX bright blood image showing a dilated RV with microaneurysms and dyskinesis of the inferior wall, acute angle, and anterior wall of the RV (arrows). (D) T1-weighted image reveals moderate subepicardial fat infiltration of the RV anterior wall, extending as "fingers" into the myocardium (arrow). Note fatty infiltration in the LV lateral wall (arrowhead). (E) RVOT bright blood image showing dyskinesis of the RVOT and RV inferior wall (arrows). (F) Horizontal long-axis bright blood image showing dyskinesis of the subtricuspid region, RV anterior wall, and RV apex (arrows). ARVD/C, arrhythmogenic right ventricular dysplasia/cardiomyopathy; CMR, cardiac magnetic resonance; RV, right ventricular; RVOT, right ventricular outflow tract. *Reproduced with permission from Ref. [7].*

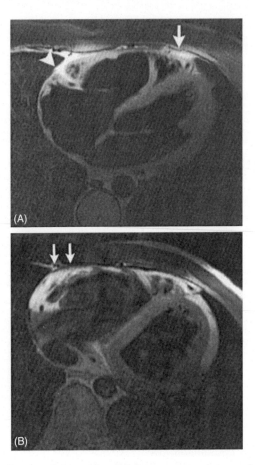

FIGURE 4.5   **Differences in the epicardial fat appearance in normal volunteer versus patient with ARVC/D.** (A) Axial black-blood image from a normal volunteer showing a clear line of demarcation between the epicardial fat and the underlying myocardium. Also note the abundance of epicardial fat in the atrioventricular groove (arrowhead) and at the apex (arrow). (B) Axial black-blood image from a patient with RV dysplasia showing lack of demarcation between epicardial fat and myocardium (arrows). *Reproduced with permission from Ref. [18].*

The role of CMR in the diagnosis of ARVC/D includes demonstration and identification of intramyocardial fat deposits, pathological myocardial wall inflammation, and fibrosis leading to a thinning and aneurysmal changes in the myocardium.

Fatty infiltration of the RV appears bright on T1-weighted sequences in an otherwise normal myocardium that has intermediate intense signal [16,17]. (Fig. 4.5). A similiar process was shown to involve the LV and is also diagnosed by CMR (Fig. 4.6).

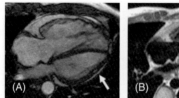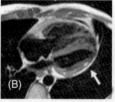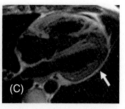

FIGURE 4.6   **Tissue characterization of the CMR allows differentiation of LV fatty infiltration in a patient with ARVC/D.** All images were obtained on the same slice position for consistency. (A) SSFP (bright blood) axial image revealing a pathological RV with the ARVC/D features as well as a small focal subepicardial signal in the distal anterolateral LV wall. (B) TSE (black blood, T2W) images show a hyperintense signal (comparable in brightness with the subcutaneous fat) in the same area of the LV. (C) Triple inversion recovery sequence (STIR) suppressing fat signal and demonstrating a hypointense signal in the distal anterolateral LV wall, thus confirming the lipid-rich composition of the LV wall lesion.

CMR PROTOCOL AND CARDIAC ANATOMY IN ARVC/D

In the next chapter, we describe in detail the CMR acquisition protocol that we utilize at the University of Arizona for patients with known or suspected ARVC/D.

The ventricular anatomical and functional assessment is usually done with T1-weighted bright blood steady-state free precession (SSFP) sequences. These sequences help assess the presence of aneurysms or dilatation of the RV. Cine sequences of the ventricle permit an assessment of the WMAs and volumetric quantification of RV function. It also helps to assess involvement of the RVOT and LV. This is particularly valuable in evaluating patients with the different phenotypes such as those with desmoplakin gene mutations where there is marked involvement of the LV with relative sparing of the RV [19].

The most important CMR finding to determine the presence of ARVC/D is assessment of the RV size (one of the major TF criteria). However, the size of the RV is not a specific sign of ARVC/D (Fig. 4.7): RV enlargement may be present in many other pathological conditions, such as pulmonary embolus, congenital diseases causing RV volume/pressure overload, RV infarction, and cor pulmonale. Some endurance athletes may have a mildly enlarged RV cavity size, without any other features of cardiac pathology.

Increased myocardial trabeculation of the RV as well as a prominent (hyperreflective) moderator band in ARVC/D may be a useful finding suggestive of the disease (Fig. 4.8). This finding is not specific for ARVC/D and may be related to the presence of noncompaction cardiomyopathy (NCCMP). However, in patients with NCCMP, the majority of the changes

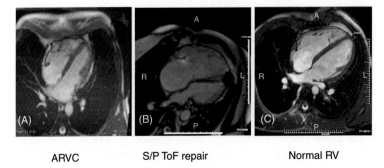

ARVC   S/P ToF repair   Normal RV

FIGURE 4.7   **Examples of the RV enlargement on CMR.** (A) The patient with advanced ARVC; (B) the patient with severe pulmonary regurgitation and s/p tetralogy of Fallot repair; (C) endurance athlete with no cardiac pathology.

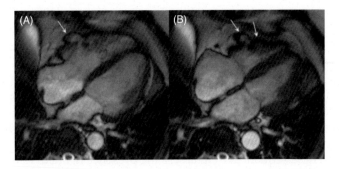

FIGURE 4.8   **Typical CMR features of the patient with ARVC/D; dilated RV, global RV systolic dysfunction; trabecular disarray, focal aneurysms of the RV free wall.** (A) Diastole; (B) Systole. *Reproduced with permission from Ref. [20].*

involve the LV, and there is a definitive decrease in the compacted myocardial thickness, predominant trabeculation of the apex and distal anterolateral segment, suggesting non-ARVC/D pathology (Fig. 4.9). In contrast, predominant RV trabeculation and preserved thickness of the LV compacted myocardium are the findings of clinical importance for the diagnosis of ARVC/D.

As previously mentioned, T1-weighted imaging sequences may help to identify fibrofatty deposition with its characteristic hyperintense signal [21]. A combination of the RV short axis and the RV axial images helps to analyze the smaller areas of fibrofatty infiltration. The presence of fat can also be assessed by inversion recovery sequences that cause homogenous fat suppression for better tissue characterization (Fig. 4.6). Identification of fibrofatty infiltration can be difficult due to the thin RV and motion artifacts of the RV free wall. Arrhythmias may decrease the

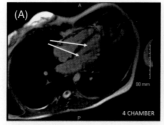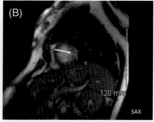

FIGURE 4.9   **CMR of the patient with NCCMP.** Pathological trabeculation of the LV apex and distal (apical) anterolateral segment is present (arrows) on these SSFP images. The compacted myocardium in the affected areas is thin, leading to a ratio of compacted to noncompacted myocardium of less than 1:2.3 in diastole. The RV is not affected. (A) 4 Chamber view; (B) Short axis view.

sensitivity of this identification. The presence of intramyocardial fat deposition is not pathognomonic for ARVC/D as it is found in normal hearts [22].

ASSESSMENT OF MYOCARDIAL FUNCTION

ARVD is characterized by infiltrative and subsequent inflammatory changes of the RV that leads to regional as well as global myocardial dysfunction. The morphological changes may not be readily apparent in the early subclinical phase. The MRI cine sequences with SSFP methodology can assist in assessing regional and global RV function. Regional WMAs can be seen in the RV free wall, outflow tract, and apex and can be homogenous or heterogeneous in distribution. As the disease progresses, CMR can identify increased RV volume (increased RVEDV) and lower RV output (RVEF) [23].

CMR also helps to differentiate ARVC/D from mimicking diseases such as sarcoidosis, idiopathic RV outflow tachycardia, myocarditis, and others [24–30].

MYOCARDIAL WALL ASSESSMENT

One of the hallmarks of ARVC/D is the development of myocardial fibro-fatty infiltration and fibrosis. The resulting scarring can be readily identified with Gd-enhanced delayed enhancement images (Fig. 4.10). In the fibrotic or scarred myocardium, there is an increase in the volume of distribution of Gd due to interstitial space expansion. This results in a hyperintense signal in the delayed enhancement images in those myocardial regions [31]. Ever since Gd enhancement sequences were

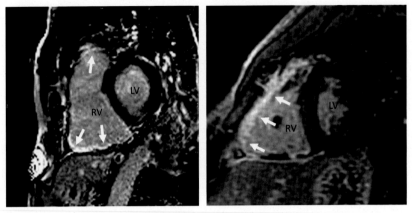

FIGURE 4.10   Evidence of the extensive RV fibrotic changes (white arrows) in patients with advanced ARVC/D on the delayed enhancement CMR imaging.

introduced, it has become helpful to evaluate the presence of scar in the myocardium. It has also been shown to correlate with histopathology and to predict future occurrence of ventricular tachycardia [6,32]. Delayed enhancement is not specific for ARVC/D and can be seen in mimics of ARVC/D including sarcoidosis, myocarditis, and even in myocardial infarction.

LIMITATIONS OF CMR

CMR has several limitations in evaluating ARVC/D. The protocols are operator dependent. Experienced CMR technicians are likely to obtain a diagnostic study compared to those with a weak knowledge of the diagnostic targets. The CMR findings that are characteristic of ARVC/D are not specific for the disease [6]. The disease process may predominately involve the LV (Fig. 4.11) [33]. In addition, there can be significant intraobserver variability between interpreting physicians [21,34]. Variability exists not only in defining the quantitative data, but even defining normal versus abnormal scans. Again, clinical experience is an important factor to accurately interpreting the scans of patients with possible ARVC/D.

Recently, the presence of the regional WMAs in healthy subjects with normal RV was confirmed by 3-T CMR imaging as compared to the standard acquisition on 1.5-T magnets. (Figs 4.12 and 4.13) [35]. The authors reported the presence of hypokinesis, dyskinesis, and bulging in 92% of the healthy patients and involvement of two or more segments in 60%.

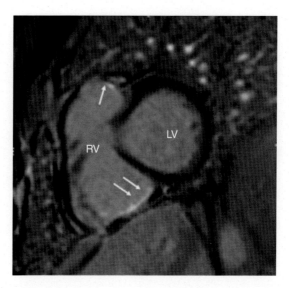

FIGURE 4.11 **Evidence of iatrogenic subendocardial scar (white arrows) due to the ablation procedure in a patient who underwent a successful catheter ablation of RV ventricular tachycardia.** LV, left ventricle; RV, right ventricle.

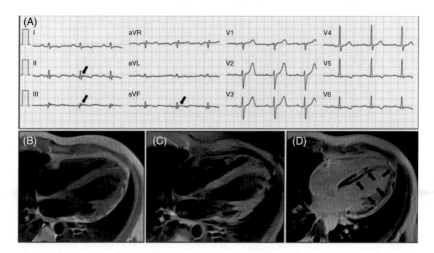

FIGURE 4.12 **Isolated arrhythmogenic LV dysplasia.** Legend from the original reference: (A) the 12-lead electrocardiogram showed sinus rhythm with epsilon waves in II, III, and aVF and T-wave inversion in inferolateral leads. (B) T1-weighted CMR images without fat suppression in four-chamber plane showed LV systolic dysfunction (LVEF = 41%) with subepicardial fat deposits in LV lateral wall apical segment (arrows). (C) T1-weighted CMR images with fat suppression confirmed findings described earlier. (D) Extensive LV intramural and subepicardial fibrosis on late-enhancement images. *Reproduced with permission from Ref. [33].*

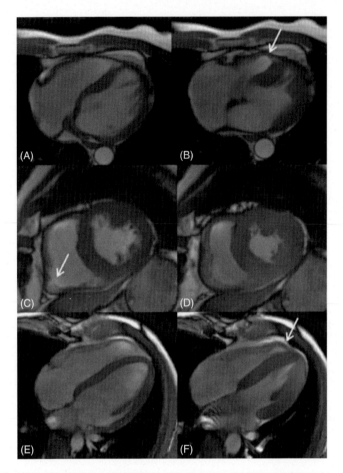

FIGURE 4.13 **RV global and regional function in healthy subjects scanned by 3-T CMR.** Legend from the original reference: Images of the RV wall motion in diastole (A, C, E) and systole (B, D, F) acquired in different planes. (A and B) Transverse view showing dyskinesia in the mediolateral segment in systole (A, white arrow). (C and D) Short-axis view with prominent bulging in the inferolateral segment in diastole (C, white arrow). (E and F) Horizontal long-axis view demonstrating dyskinesia in systole in apicolateral segment (F, white arrow). *Reproduced with permission from Ref. [35].*

Importantly, the size of the focal WMA was small and no WMA were found in the basal region of the RV. In this study, normal patients had regional WMAs predominantly in the apicolateral and mediolateral segments with a close spatial relation to the moderator band (79.3%) and trabecular muscle (19.3%) [35]. This observation may help to distinguish a "typical" pattern of a basal location of the abnormal focal RV geometry in ARVC/D patients.

Despite its limitations, CMR has an exceptionally important role in the evaluation of patients with known or suspected ARVC/D and is an excellent diagnostic test.

References

[1] Diamond GA, Forrester JS. Analysis of probability as an aid in the clinical diagnosis of coronary-artery disease. N Engl J Med 1979;300:1350–8.

[2] Staniloff HM, Diamond GA, Freeman MR, Berman DS, Forrester JS. Simplified application of Bayesian analysis to multiple cardiologic tests. Clin Cardiol 1982;5:630–6.

[3] Abidov A, Hachamovitch R, Hayes SW, et al. Are shades of gray prognostically useful in reporting myocardial perfusion single-photon emission computed tomography? Circ Cardiovasc Imaging 2009;2:290–8.

[4] Leipsic J, Abbara S, Achenbach S, et al. SCCT guidelines for the interpretation and reporting of coronary CT angiography: a report of the Society of Cardiovascular Computed Tomography Guidelines Committee. J Cardiovasc Comput Tomogr 2014;8:342–58.

[5] Marcus FI, McKenna WJ, Sherrill D, et al. Diagnosis of arrhythmogenic right ventricular cardiomyopathy/dysplasia: proposed modification of the task force criteria. Circulation 2010;121:1533–41.

[6] Sen-Chowdhry S, Prasad SK, Syrris P, et al. Cardiovascular magnetic resonance in arrhythmogenic right ventricular cardiomyopathy revisited: comparison with task force criteria and genotype. J Am Coll Cardiol 2006;48:2132–40.

[7] te Riele AS, Bhonsale A, James CA, et al. Incremental value of cardiac magnetic resonance imaging in arrhythmic risk stratification of arrhythmogenic right ventricular dysplasia/cardiomyopathy-associated desmosomal mutation carriers. J Am Coll Cardiol 2013;62:1761–9.

[8] Deac M, Alpendurada F, Fanaie F, et al. Prognostic value of cardiovascular magnetic resonance in patients with suspected arrhythmogenic right ventricular cardiomyopathy. Int J Cardiol 2013;168:3514–21.

[9] Keller DI, Osswald S, Bremerich J, et al. Arrhythmogenic right ventricular cardiomyopathy: diagnostic and prognostic value of the cardiac MRI in relation to arrhythmia-free survival. Int J Cardiovasc Imaging 2003;19:537–43. discussion 45–47.

[10] Vermes E, Strohm O, Otmani A, Childs H, Duff H, Friedrich MG. Impact of the revision of arrhythmogenic right ventricular cardiomyopathy/dysplasia task force criteria on its prevalence by CMR criteria. JACC Cardiovasc Imaging 2011;4:282–7.

[11] Femia G, Hsu C, Singarayar S, et al. Impact of new task force criteria in the diagnosis of arrhythmogenic right ventricular cardiomyopathy. Int J Cardiol 2014;171:179–83.

[12] Herren T, Gerber PA, Duru F. Arrhythmogenic right ventricular cardiomyopathy/dysplasia: a not so rare "disease of the desmosome" with multiple clinical presentations. Clin Res Cardiol 2009;98:141–58.

[13] Khoo NS, Young A, Occleshaw C, Cowan B, Zeng IS, Gentles TL. Assessments of right ventricular volume and function using three-dimensional echocardiography in older children and adults with congenital heart disease: comparison with cardiac magnetic resonance imaging. J Am Soc Echocardiogr 2009;22:1279–88.

[14] Tada H, Shimizu W, Ohe T, et al. Usefulness of electron-beam computed tomography in arrhythmogenic right ventricular dysplasia. Relationship to electrophysiological abnormalities and left ventricular involvement. Circulation 1996;94:437–44.

[15] Kayser HW, van der Wall EE, Sivananthan MU, Plein S, Bloomer TN, de Roos A. Diagnosis of arrhythmogenic right ventricular dysplasia: a review. Radiographics 2002;22:639–48. discussion 49–50.

[16] Auffermann W, Wichter T, Breithardt G, Joachimsen K, Peters PE. Arrhythmogenic right ventricular disease: MR imaging vs angiography. AJR Am J Roentgenol 1993;161:549–55.

[17] White RD, Trohman RG, Flamm SD, et al. Right ventricular arrhythmia in the absence of arrhythmogenic dysplasia: MR imaging of myocardial abnormalities. Radiology 1998;207:743–51.

[18] Tandri H, Bomma C, Calkins H, Bluemke DA. Magnetic resonance and computed tomography imaging of arrhythmogenic right ventricular dysplasia. J Magn Reson Imaging 2004;19:848–58.

[19] Sen-Chowdhry S, Syrris P, Prasad SK, et al. Left-dominant arrhythmogenic cardiomyopathy: an under-recognized clinical entity. J Am Coll Cardiol 2008;52:2175–87.

[20] Karamitsos TD, Francis JM, Myerson S, Selvanayagam JB, Neubauer S. The role of cardiovascular magnetic resonance imaging in heart failure. J Am Coll Cardiol 2009;54: 1407–24.

[21] Bluemke DA, Krupinski EA, Ovitt T, et al. MR Imaging of arrhythmogenic right ventricular cardiomyopathy: morphologic findings and interobserver reliability. Cardiology 2003;99:153–62.

[22] Burke AP, Farb A, Tashko G, Virmani R. Arrhythmogenic right ventricular cardiomyopathy and fatty replacement of the right ventricular myocardium: are they different diseases? Circulation 1998;97:1571–80.

[23] Ferrari VA, Scott CH. Arrhythmogenic right ventricular cardiomyopathy: time for a new look. J Cardiovasc Electrophysiol 2003;14:483–4.

[24] Barbou F, Lahutte M, Gontier E. Cardiac sarcoidosis presenting as an arrhythmogenic right ventricular cardiomyopathy. Heart 2012;98:1753–4.

[25] Chia PL, Subbiah RN, Kuchar D, Walker B. Cardiac sarcoidosis masquerading as arrhythmogenic right ventricular cardiomyopathy. Heart Lung Circ 2012;21:42–5.

[26] Corrado D, Thiene G. Cardiac sarcoidosis mimicking arrhythmogenic right ventricular cardiomyopathy/dysplasia: the renaissance of endomyocardial biopsy? J Cardiovasc Electrophysiol 2009;20:477–9.

[27] Tandri H, Bluemke DA, Ferrari VA, et al. Findings on magnetic resonance imaging of idiopathic right ventricular outflow tachycardia. Am J Cardiol 2004;94:1441–5.

[28] Chimenti C, Calabrese F, Thiene G, Pieroni M, Maseri A, Frustaci A. Inflammatory left ventricular microaneurysms as a cause of apparently idiopathic ventricular tachyarrhythmias. Circulation 2001;104:168–73.

[29] Christensen AH, Svendsen JH. Myocarditis mimicking arrhythmogenic right ventricular cardiomyopathy. J Am Coll Cardiol 2009;54:663–4. author reply 5–6.

[30] Divakara Menon SM, Ganame J, Lira Lea Plaza G, et al. Right ventricular tachycardia: common presentation versus common disease. Circulation 2013;128:e85–7.

[31] Tandri H, Saranathan M, Rodriguez ER, et al. Noninvasive detection of myocardial fibrosis in arrhythmogenic right ventricular cardiomyopathy using delayed-enhancement magnetic resonance imaging. J Am Coll Cardiol 2005;45:98–103.

[32] Tandri H, Calkins H, Nasir K, et al. Magnetic resonance imaging findings in patients meeting task force criteria for arrhythmogenic right ventricular dysplasia. J Cardiovasc Electrophysiol 2003;14:476–82.

[33] Raza A, Waleed M, Balerdi M, Bragadeesh T, Clark AL. Idiopathic left ventricular apical hypoplasia. BMJ Case Rep 2014;2014.

[34] Tandri H, Calkins H, Marcus FI. Controversial role of magnetic resonance imaging in the diagnosis of arrhythmogenic right ventricular dysplasia. Am J Cardiol 2003;92:649.

[35] Quick S, Speiser U, Kury K, Schoen S, Ibrahim K, Strasser R. Evaluation and classification of right ventricular wall motion abnormalities in healthy subjects by 3-tesla cardiovascular magnetic resonance imaging. Neth Heart J 2015;23:64–9.

5

Current Cardiac MRI Protocols for Known and Suspected ARVC/D

Aiden Abidov, Isabel B. Oliva

**Department of Medicine/Division of Cardiology
and Department of Medical Imaging,
University of Arizona, Tucson, AZ, USA**

Cardiac MRI (CMR) is a versatile imaging method capable of successfully resolving diagnostic challenges encountered by the clinicians working with known or suspected ARVC/D patients.

The modified 2010 ARVC/D Task Force criteria [1] consider echocardiography and CMR as the only imaging modalities of the disease. For both modalities, the guidelines identify the structure and morphological changes (the RV size as well as the global and regional systolic RV dysfunction) as target parameters to establish the diagnosis of ARVC/D (Table 5.1; Fig. 5.1).

These experts evaluated these imaging variables based on the diagnostic certainty, repeatability, and reproducibility of the parameters. These are important considerations since the uniformity of diagnosis based on reliable and reproducible data creates a basis for standardization as well as achievement of a greater uniformity when the data are used in multicenter registries or in clinical centers with multiple readers. The experts did not include tissue characterization (fatty infiltration or RV myocardial fibrosis) among the diagnostic CMR criteria of the disease based on the same concern of reproducibility and reliability [1].

The diagnostic performance of CMR in patients with ARVC/D has been impressive, even in early publications on this topic [2]. CMR maintains its role as an effective diagnostic tool for this entity. As an example, within the population of patients from families who were identified

TABLE 5.1   Imaging Criteria of ARVC/D Included in the 2010 Modified Task Force
Criteria

1. Global or regional dysfunction and structural alterations

Major

  2D Echo Criteria

Regional RV akinesia, dyskinesia, or aneurysm and one of the following measured at
end diastole
- PLAX RVOT ≥ 32 mm (PLAX/BSA ≥ 19 mm/m²), or
- PSAX RVOT ≥ 36 mm (PSAX/BSA ≥ 21 mm/m²), or
- Fractional area change ≤ 33%

  CMR criteria

Regional RV akinesia or dyskinesia or dyssynchronous RV contraction and one of the
following
- RV EDV/BSA ≥ 110 mL/m² (male) or ≥100 mL/m² (female)
- RV ejection fraction ≤ 40%

  RV angiography criteria

Regional RV akinesia, dyskinesia, or aneurysm

Minor

  2D Echo criteria

Regional RV akinesia or dyskinesia or dyssynchronous RV contraction and one of the
following measured at end diastole
- PLAX RVOT ≥ 29 to < 32 mm (PLW/BSA ≥ 16 to < 19 mm/m²), or
- PSAX RVOT ≥ 32 to < 36 mm (PSAX/BSA ≥ 18 to < 21 mm/m²), or
- Fractional area change > 33% ≤ 40%

  CMR criteria

Regional RV akinesia or dyskinesia or dyssynchronous RV contraction and one of the
following
- RV EDV/BSA ≥ 100 to 110 mL/m² (male) or ≥90 to 100 mL/m² (female)
- RV ejection fraction >40 to ≤45%

Modified with permission from Ref. [5].

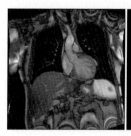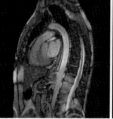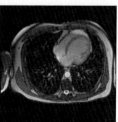

FIGURE 5.1   **Cardiac localizers.**

as pathogenic ARVC/D-associated desmosomal mutations carriers [3], CMR findings identified by the ARVD/C TFC panel demonstrated over-all good performance (Table 5.2): the majority of patients with abnormal CMR results fulfilled the major TFC (81%). The subset of patients who presented with arrhythmias was significantly more likely to fulfill the TFC for CMR (64%) when compared to asymptomatic patients (15%) (p value < 0.001).

Overall, CMR is an excellent diagnostic tool, but it requires skill and expertise, and it appears to have a "learning curve" for both the physician and technician. It requires close collaboration and communication with the entire clinical team, starting with the referring physicians and their understanding of appropriateness of the test they have ordered for their patient.

A dedicated CMR protocol for image acquisition is probably the most important part of the diagnostic triangle ("appropriate order – right image – correct interpretation") that we have described earlier.

Different CMR scanner vendors provide sequence terminologies and methodologies specific for their software and hardware, which adds additional complexity in the image acquisition process.

Our current University of Arizona's CMR protocol for ARVC/D is described in the next chapter. We utilize Siemens MRI scanners, but in general, all the manufactures provide robust and well-balanced methodology for CMR acquisition and thus, our protocol can be used in any scanner.

We will use general terminology to describe the CMR sequences. Our predominant sequence on 1.5-Tesla magnet for assessment of the global and regional RV and LV function is a balanced steady state free precession sequence, or bSSFP. Rarely, we may need to utilize the gradient echo (GRE)–based cine images, mostly when significant valvular regurgitation is suggested or when the patient has a pacemaker/ICD (see Chapter 9).

Our ARVC/D protocol is fairly standardized and provides structural and functional data for both ventricles since LV involvement is an important clinical and prognostic factor in this disorder. In addition, there are many clinical conditions that may mimic ARVC/D and thus, the CMR data should have a sufficient amount of information to successfully differentiate these disorders. For these reasons, our protocol includes additional sequences such as the rest perfusion and phase-contrast data. The protocol may be further adjusted and customized by the physician after reviewing initial images on the scanner. We strongly recommend physician review while the patient is still on the scanner, since we frequently see findings requiring change or addition of new sequences and views. Every possible effort should be made in order to obtain the most informative and effective CMR study

TABLE 5.2   CMR Parameters in Symptomatic and Asymptomatic ARVC/D associated Desmosomal Mutation Gene Carriers

Variable	No arrhythmic event (*n* = 58)	Arrhythmic event (*n* = 11)	*p* value
Men	21 (36%)	8 (73%)	0.024
Proband	7 (12%)	8 (73%)	<0.001
Symptomatic at presentation	15 (26%)	7 (64%)	0.014
Syncope	7 (12%)	2 (18%)	NS
Presyncope	4 (7%)	2 (18%)	NS
Palpitations	10 (17%)	6 (55%)	0.007
Chest pain	2 (3%)	1 (9%)	NS
Abnormal results on ECG	29 (50%)	10 (91%)	0.012
Negative T waves in leads V_1 and V_2	6 (10%)	1 (9%)	NS
Negative T waves in leads V_1 to V_3 or beyond	19 (33%)	9 (82%)	0.002
Epsilon waves	0 (0%)	1 (9%)	0.021
Terminal activation duration $\geq$55 ms	9 (16%)	4 (36%)	NS
Abnormal results on Holter monitoring*	10/49 (20%)	4/5 (80%)	0.004
>500 PVCs/24 h	9/49 (18%)	4/5 (80%)	0.002
Nonsustained VT recorded	6/49 (12%)	3/5 (60%)	0.006
Fulfillment of TFC for CMR	10 (17%)	11 (100%)	<0.001
Major TFC	6 (10%)	11 (100%)	<0.001
Minor TPC	4 (7%)	0 (0%)	NS
RV EDV/BSA (mL/m²)	80.9 ± 19.2	109.9 ± 20.7	<0.001
LV EDV/BSA (mL/m²)	78.6 ± 13.9	87.7 ± 17.6	NS
RV EF (%)	49.3 ± 9.2	32.6 ± 6.8	<0.001
LV EF (%)	55.2 ± 6.0	45.8 ± 4.9	<0.001
RV wall motion abnormalities	14 (24%)	11 (100%)	<0.001
RV fatty infiltration	7 (12%)	2 (20%)	NS
RV delayed enhancement**	0/52 (0%)	2/9 (22%)	0.001
LV wall motion abnormalities	4 (7%)	2 (18%)	NS
LV fatty infiltration	7 (12%)	6 (55%)	0.001
LV delayed enhancement**	5/52 (10%)	2/9 (22%)	NS

TABLE 5.2   CMR Parameters in Symptomatic and Asymptomatic ARVC/D associated Desmosomal Mutation Gene Carriers (*cont.*)

Variable	No arrhythmic event (*n* = 58)	Arrhythmic event (*n* = 11)	*p* value
RV involvement only	8 (14%)	4 (36%)	NS
LV involvement only	3 (5%)	0 (0%)	NS
Biventricular involvement	8 (14%)	7 (64%)	<0.001

Values are *n* (%) or mean ± SD.
54 patients underwent Holter monitoring.
**For 61 patients, delayed enhancement images were available.*
Reproduced with permission from Ref. [3].

to aid in establishing the right diagnosis for every patient referred to the imaging lab.

THE UNIVERSITY OF ARIZONA ARVC/D CMR PROTOCOL

1. *Cardiac localizers* (Fig. 5.1): A few images are acquired through the chest in the axial, sagittal, and coronal planes using single-shot fast spin echo technique. These images provide a general assessment of the intrathoracic anatomy and location of the heart. They are essential for the planning of the dedicated cardiac views.
2. *Dark blood imaging* (Figs 5.2 and 5.3): Half-Fourier acquisition single-shot turbo spin-echo (HASTE) axial views of the chest to assess anatomy and aid in the planning of follow-up images. This acquisition is planned from the three-plane localizer images. The

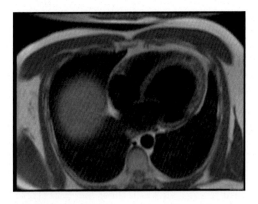

FIGURE 5.2   **Axial HASTE images.**

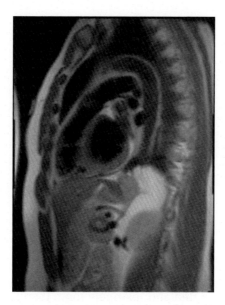

FIGURE 5.3 **HASTE sagittal view centered on the right ventricular outflow tract (RVOT), planned from the axial black blood images with the line of planning positioned parallel to the main pulmonary artery and passing through the RVOT.** These images can be used to plan dedicated RV views in the axial plane to assess RV free wall motion and characterize tissue (fat deposition).

blood is nulled by double inversion recovery (DIR) technique. Stagnant blood within the LV trabeculations will demonstrate high signal and should not be confused with subendocardial edema. Single-shot technique allows acquisition of multiple images in a single breath hold. DIR images are always gated to the ECG to remove cardiac motion.

In some ARVC/D imaging centers, the breathhold DIR fast SE (DIR FSE) sequences is preferred and HASTE technique is not recommended due to the potential of cardiac motion artifact [4]. However, in our center we have consistently adequate clinical performance of the HASTE sequences; in case of suspected positive findings on HASTE, we may switch to breathhold acquisition for clarity. In this regard, it is important that the CMR physicians review the initial images prior to the contrast injection to establish the need for additional sequences before the gadolinium exposure.

3. *Vertical long axis (two-chamber) SSFP cine* (Fig. 5.4): Bright blood technique that can be acquired using either a balanced steady-state free precession (bSSFP) technique or the conventional GRE sequence. The use of SSFP is preferable due to its higher contrast

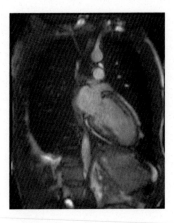

FIGURE 5.4   **Two-chamber SSFP cine images.**

resolution and signal-to-noise ratio. GRE has lower contrast-to-noise ratio but should be used in patients with pacemaker/ICDs. Images are ECG gated and relatively long; therefore, usually one slice is acquired per breath hold. The line of planning should pass through the center of the mitral valve, including the left atrium (LA) and LV apex.

SSFP is a GRE sequence that acquires images by repeatedly applying a gradient with low flip angle and short TR that leads to a steady state of longitudinal magnetization. The SSFP images are strongly susceptible metal artifact and are heavily dependent on homogeneity of the magnetic field.

4. *Short-axis (SAX) Scout* (Fig. 5.5): Cine bSSFP technique acquired through the base of the heart and midventricle and prescribed from

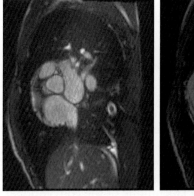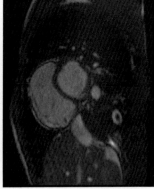

FIGURE 5.5   **SAX SSFP images.**

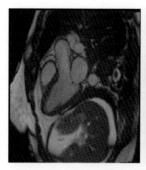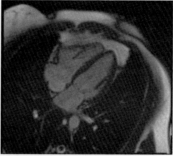

FIGURE 5.6   **Three- and four-chamber SSFP images.**

the two-chamber view. These images are vital for the planning of the three- and four-chamber views. The initial acquisition images the base of the LV and aortic root and the second plane goes through both ventricles.

5. *Three- and four-chamber dedicated LV SSFP cine images* (Fig. 5.6): Cine bright blood images of the heart centered on the mitral valve and planned from the combination of the SAX scout and two-chamber views. The three-chamber view is acquired when positioning the line of planning through the center of the mitral valve and middle of the LA and LV on the two-chamber view, and through the left ventricular outflow tract (LVOT) and aortic root on the first image of the SAX scout acquisition. The four-chamber view is acquired when positioning the line of planning through the mitral valve and middle of the LA and LV on the two-chamber view, and through the mid-LV cavity on the second image of the SAX scout acquisition. It is important to rotate the line toward the largest position of the RV.

6. *Dedicated RV cine views* (Fig. 5.7): We recommend obtaining dedicated RV inflow and outflow cine images since these provide optimal visualization of the triangle of dysplasia, tricuspid valve, and pulmonic valve. These are planned from the sagittal black blood images and are intended to evaluate wall motion of the RV free wall. Motion of the RVOT can also be assessed in this plane.

7. *Axial cine view through the RV* (Fig. 5.8): Cine bSSFP images through the RV on axial plane.

8. *T1-weighted images of the RV* (Fig. 5.9): Black blood T1-weighted spin echo images acquired through the RV to assess for fat deposition in the anterior and free walls.

We recommend two separate acquisitions: the initial acquisition is made through the RV on the axial plane to assess the anterior RV wall. The slices should be exactly the same as the axial cine RV images

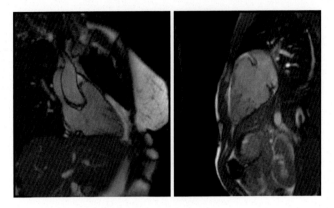

FIGURE 5.7   Dedicated RV cine views.

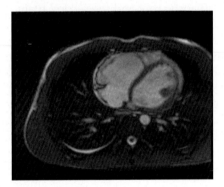

FIGURE 5.8   Axial RV-centered cine images.

so one can correlate the areas of fat deposition with abnormal wall motion. The second acquisition is done in the coronal or sagittal plane to analyze the inferior RV wall.

If a hyperintense focus is seen within the RV myocardium, this sequence is repeated with the addition of chemical fat saturation to confirm fat deposition.

9. *Rest perfusion* (Fig. 5.10): First pass rest perfusion images are acquired in the SAX (three slices, basal – mid – apical SAX) and four-chamber (one slice) planes as part of the routine examination. This is a T1-weighted sequence with images acquired immediately after intravenous administration of the gadolinium-based contrast. We use a 0.05 mmol/kg dose injected at 3 cm^3/s. After image acquisition is completed, the patient receives an additional half dose of contrast to allow appropriate delayed enhancement imaging.

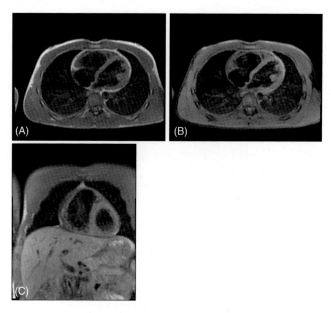

FIGURE 5.9   **Axial Black blood T1 WI SE without (A) and with (B) fat saturation. Coronal Black blood T1 WI SE without fat saturation (C).**

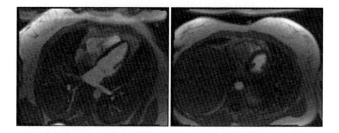

FIGURE 5.10   **Rest perfusion images.**

10. *SAX SSFP cine stack* (Fig. 5.11): Cine bright blood images through the ventricles. This sequence is planned from the four- and two-chamber views to insure that both ventricles are included in their entirety. A planning line is placed perpendicular to the interventricular septum in the four-chamber view and parallel to the mitral valve in the two-chamber view. These images are required to calculate ventricular volume and ejection fraction, both of which are included in the revised TFC for ARVC/D. They are also used to assess both left and RV wall motion as well as the function of the mitral and tricuspid valves.

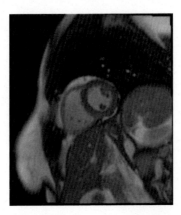

FIGURE 5.11   **SAX SSFP cine images.**

FIGURE 5.12   **LVOT cine images.**

11. *LVOT SSFP cine images* (Fig. 5.12): Cine bright blood image through the ascending thoracic aorta is planned from the three-chamber. This sequence is used in conjunction with the three-chamber view to evaluate the aortic root and aortic valves. Both these planes are also used to plan the cine images of the aortic valve and phase-contrast imaging through the ascending aorta.

12. *Aortic valve SSFP cine* (Fig. 5.13): Cine bright blood image through the aortic valve planned from the LVOT and three-chamber cine images. Usually three slices are acquired (at, below, and above the valve). These images are used to accurately measure the dimensions of the aortic sinuses, give an appropriate view of the aortic valve leaflets, and assess for dissection flaps.

13. *TI scout* (Fig. 5.14): Single ECG-gated slice through the middle of the left ventricle acquired 9–10 min after intravenous administration of

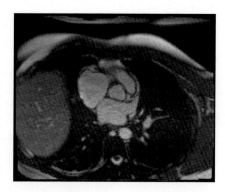

FIGURE 5.13   **Aortic valve cine images.**

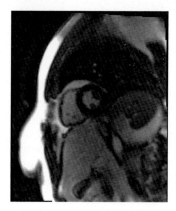

FIGURE 5.14   **TI scout.**

contrast. This image is planned from the four-chamber view and is used to determine the appropriate time of inversion for the delayed enhancement images.

14. *Myocardial delayed enhancement (MDE)* (Fig. 5.15): ECG-gated image through the heart is obtained by copying the position of the previously acquired SAX stack as well as two-, three-, and four-chamber planes. The MDE is a T1-weighted GRE sequence that uses an inversion recovery pulse to null the normal myocardium. The appropriate time of inversion is copied from the TI scout. When evaluating patients with known/suspected ARVC/D we also include stack images of the RV to evaluate the anterior and inferior walls of the RV. The TI for the RV myocardium is generally 50 ms shorter than that for the LV myocardium but

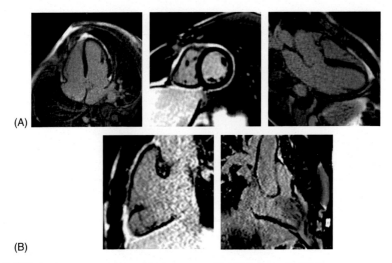

(A)

(B)

FIGURE 5.15   (A) LV dedicated MDE images. (B) RV dedicated MDE images.

an RV dedicated TI scout can be performed if needed. We use
a single-shot technique for acquisition of seven to eight slices
in one breath hold. This technique can also be performed with
free breathing due to its inherent fast time of acquisition. The
segmented technique can be used if indeterminate areas of
delayed enhancement are seen within the myocardium. This
technique has a longer acquisition time (one slice per breath
hold) but provides images that have higher contrast-to-noise
ratio and spatial resolution. Both single shot and segmented MDE
sequences provide magnitude and phase sensitivity images for
each slice.

15. *Phase-contrast (PC) imaging* (Fig. 5.16): GRE sequence is used to
measure blood flow, direction, and velocity by measuring the
phase shift of the protons in the blood when moving through
a magnetic field. Quantitated peak systolic velocity is used to
calculate pressure gradient, which is essential when evaluating
patients with suspected/known valvular stenosis. This sequence
is planned through the mid ascending aorta and main pulmonary
artery. The three-chamber and LVOT images are used to plane the
PC through the aorta, and the dedicated RVOT images are used
to plane the PC through the main pulmonary artery. Forward
volumes are compared to the stroke volumes derived from the cine
SAX stack sequence. Blood volumes and direction are also used

to estimate shunt fraction (Qp:Qs) and assess severity of valvular insufficiency.

16. *Optional*: if anomalous coronaries are suspected, the navigator sequence for visualization of the coronary arteries should be considered.

Table 5.3 summarizes technical details of the acquisition for the CMR sequences comprising the University of Arizona's ARVC/D protocol.

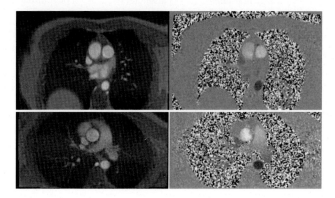

FIGURE 5.16   **Phase-contrast images of the pulmonary artery (top row) and the aorta (bottom row).**

TABLE 5.3   Technical Details for the Main CMR Sequences of the University of Arizona ARVC/D Protocol

Sequences	TE (ms)	TR (ms)	Segments	Flip angle	Slice thickness (mm)	Slice gap (%)
Localizers	1.13	288.36	96	80	7	100
Ax SSFP	1.22	236.71	13	80	7	0
HASTE	46	779	68	160	8	25
SSFP cines	1.16	35.88	68	80	6	20
T1 WI for fat	24	700	15	160	5	10
Rest perfusion	1.17	170.74	55	12	8	100
TI scout	1.11	26.30	10	30	8	20
Single-shot DE	1.14	700	97	40	8	10
Segmented DE	3.24	700	20	25	8	10
Phase-contrast imaging	3.04	42	4	20	6	NA

Our colleagues from Johns Hopkins University utilize a similar protocol [5] (Table 5.4). The main difference is the use of high-resolution dark blood sequences instead of HASTE in their protocol.

For patients with suspected ARVC/D, we recommend acquiring axial cine SSFP images to evaluate wall motion of the RV anterior wall;

TABLE 5.4   Johns Hopkins University CMR Protocol

Sequence	Imaging plane	Parameters	Comments
Double inversion recovery TSE/FSE **1.** Axial: with and without fat suppression **2.** Short axis: without fat suppression	**1.** Axial: obtain ~6–8 images centered on the left/right ventricle **2.** Short axis: obtain ~6–8 images centered on the left ventricle	TR = 2 R–R intervals, TE = 5 ms (minimum–full) (GE), TE = 30 ms (Siemens) slice thickness = 5 mm, interslice gap = 5 mm, and field of view (FOV) = 28–34 cm ETL 16–24	This sequence provides optimal tissue characterization of the RV free wall. Prescribe from the pulmonary artery to the diaphragm. Fat suppression improves reader confidence in diagnosis of RV fat infiltration.
SSFP bright blood cine images	Axial, four chamber and short axis. RV three chamber (optional)	TR/TE minimum, flip angle = 45–70°, slice thickness = 8 mm, interslice gap = 2 mm FOV = 36–40 cm, 16–20 views per segment. Parallel imaging n = 2 is desirable	Axial images are best to assess RV wall motion. RV quantitative analysis is performed on the short axis cine images.
Gadolinium is administered according to institutional protocol (usually 0.15–0.2 mmol/kg)			
TI scout	Four chamber		TI scout sequences or trial TI times to suppress normal myocardium for the right inversion time.
Delayed gadolinium imaging (phase sensitive inversion)	Axial, short axis, four chamber, and vertical	TR/TE per manufacturer recommendations flip angle = 20–25°, slice thickness = 8 mm	PSIR is more robust and independent of TI time. Optimal for imaging fibrosis.

Reproduced with permission from the Ref. [5].

usually we obtain 10 slices with a 20% gap. Especially important is a clear visualization of the subtricuspid area, not easily visible in other cardiac dedicated views.

We recommend obtaining the same axial images using T1-weighted black blood HASTE to look for fatty deposition in the RV anterior wall and in the LV. In addition to the axial black blood imaging (HASTE), we also obtain a coronal black blood HASTE to evaluate the RV inferior wall.

The perfusion images have a magnetization preparation pulse that is nonselective with a TI of 100 ms. DCE images also require a magnetization preparation pulse, which is nonselective.

The TI for the RV myocardium is about 50 ms less than the TI for the LV. Therefore, we can either scan the SAX DCE twice after adjusting the TI for the RV (subtracting 50 ms from that for the LV) or obtain another TI scout specifically for the RV. We use the first approach when we are not able to obtain adequate nulling of the RV myocardium on the initial SAX MDE that was aimed for the LV myocardium.

All sequences use [6] generalized autocalibrating partially parallel acquisitions (GRAPPA) with an acceleration factor of 2. All sequences are performed during breathhold and the number of concatenations depends on the patient's heart rate (HR). A faster HR allows acquisition of more images during the same breath hold. Images are acquired by the RR interval. Therefore, a faster HR means more RR intervals in a given time when compared to slower HRs so we can decrease the number of concatenations. Exceptions to this rule are the black blood images, since they need a preparation pulse. We can only acquire one slice per breath hold, regardless of the patient's HR.

In summary, after obtaining the CMR data, we should be able to answer the following three questions with a high level of certainty (Fig. 5.17):

1. Is the RV dilated? If yes – we should be able to provide a quantitative indexed value of the RVEDV/BSA.
2. Is there any regional RV dysfunction (akinesia, dyskinesia, or dyssynchronous RV contraction)?
3. Is the RVEF decreased?

The answer to these questions potentially covers all the information needed in regard to the CMR TFC. Additional CMR findings (fibro-fatty infiltration, significant RV fibrosis, increased RV trabeculation and disarray) as well as novel parameters (such as RV and biventricular strain) may further enhance diagnostic information and increase overall certainty for the diagnosis; however, this information is not yet included in the current ARVC/D criteria.

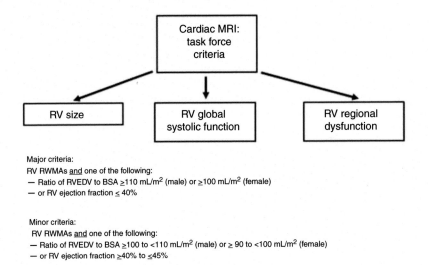

Major criteria:
RV RWMAs <u>and</u> one of the following:
— Ratio of RVEDV to BSA $\geq$110 mL/m^2 (male) or $\geq$100 mL/m^2 (female)
— or RV ejection fraction $\leq$ 40%

Minor criteria:
RV RWMAs <u>and</u> one of the following:
— Ratio of RVEDV to BSA $\geq$100 to <110 mL/m^2 (male) or $\geq$ 90 to <100 mL/m^2 (female)
— or RV ejection fraction $\geq$40% to $\leq$45%

FIGURE 5.17   **Target CMR criteria for establishing the diagnosis of ARVC/D.** RWMAs, regional wall motion abnormalities.

METHODOLOGICAL CONSIDERATIONS AND FUTURE DEVELOPMENTS TO ASSESS RV SIZE AND FUNCTION

Current RV volumetric assessment is similar for all the commercially available software and is based on Simpson's rule. It has all the flaws due to multiple geometric assumptions and simplification of the complex RV geometry. The major problem with the RV volume assessment is that the same principles used for the LV are "forcefully" applied to the RV, despite the fact that the RV has a distinctly different geometry. A challenging part of the measurement methodology for the RV is the infundibular segment. Future CMR developments will undoubtedly include, a significant level of standardization and computer assessment of the RV function and morphology. In current practice, we observe a wide variability in the assessment of the RVEF from different labs. There are a few relatively simple rules in order to maintain good repeatability and reproducibility of the results. These rules were recently summarized by Tavano et al. [7] as follows (Fig. 5.18):

1. When Simpson's disk method is utilized, it is important to include the RV infundibulum in the measurement of the RV volumes, as it can account for 25–30% of RV volume.

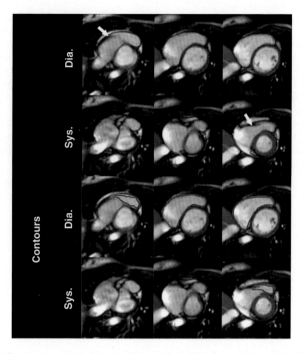

FIGURE 5.18   **Recommended methodology of the RV contouring for the RVEF calculation.** The end-systolic (Sys.) and end-diastolic (Dia.) frames are identified and compared. It is clearly shown that the most basal RV slice differs in systole and diastole. The infundibulum is included in the measurement. The infundibulum and the right atrium are separated by the atrioventricular groove (arrow). *Reproduced with permission from Ref. [7].*

2. The most basal level of the RV should be carefully identified using the SAX and LAX views. The AV groove should be used to separate the RV from the right atrium.
3. Endocardial contouring of the ventricle should be performed with inclusion of the RV trabeculae in the RV cavity.

References

[1] Marcus FI, McKenna WJ, Sherrill D, Basso C, Bauce B, Bluemke DA, Calkins H, Corrado D, Cox MG, Daubert JP, Fontaine G, Gear K, Hauer R, Nava A, Picard MH, Protonotarios N, Saffitz JE, Sanborn DM, Steinberg JS, Tandri H, Thiene G, Towbin JA, Tsatsopoulou A, Wichter T, Zareba W. Diagnosis of arrhythmogenic right ventricular cardiomyopathy/dysplasia: proposed modification of the task force criteria. Circulation 2010;121:1533–41.
[2] Casolo GC, Poggesi L, Boddi M, Fazi A, Bartolozzi C, Lizzadro G, Dabizzi RP. ECG-gated magnetic resonance imaging in right ventricular dysplasia. Am Heart J 1987;113:1245–8.
[3] te Riele AS, Bhonsale A, James CA, Rastegar N, Murray B, Burt JR, Tichnell C, Madhavan S, Judge DP, Bluemke DA, Zimmerman SL, Kamel IR, Calkins H, Tandri H. Incremental value of cardiac magnetic resonance imaging in arrhythmic risk stratification of

arrhythmogenic right ventricular dysplasia/cardiomyopathy-associated desmosomal mutation carriers. J Am Coll Cardiol 2013;62:1761–9.

[4] Tandri H, Bomma C, Calkins H, Bluemke DA. Magnetic resonance and computed tomography imaging of arrhythmogenic right ventricular dysplasia. J Magn Reson Imaging 2004;19:848–58.

[5] te Riele AS, Tandri H, Bluemke DA. Arrhythmogenic right ventricular cardiomyopathy (ARVC): cardiovascular magnetic resonance update. J Cardiovasc Magn Reson 2014;16:50.

[6] Blaimer M, Breuer F, Mueller M, Heidemann RM, Griswold MA, Jakob PM. Smash, sense, pils, grappa: how to choose the optimal method. Top Magn Reson Imag 2004;15:223–36.

[7] Tavano A, Maurel B, Gaubert JY, Varoquaux A, Cassagneau P, Vidal V, Bartoli JM, Moulin G, Jacquier A. MR imaging of arrhythmogenic right ventricular dysplasia: what the radiologist needs to know. Diagn Interv Imaging 2015;96:449–60.

Association of Phenotype and Genotype in the Diagnosis and Prognosis of ARVC/D in the Adult Population

Amit Patel*, Luisa Mestroni*, Frank I. Marcus**

*Cardiovascular Institute, University of Colorado Anschutz
Medical Campus, Aurora, CO, USA
**Department of Medicine/Division of Cardiology
and Department of Medical Imaging,
University of Arizona, Tucson, AZ, USA

INTRODUCTION

Arrhythmogenic right ventricular cardiomyopathy/dysplasia (ARVC/D) is a disease of the right ventricle (RV) in which progressive fibro-fatty tissue replaces normal myocardium and leads to an arrhythmogenic cardiac substrate, progressive dysfunction of the RV, and less commonly of the left ventricle (LV) [1]. Typically the disease is inherited in an autosomal dominant fashion, although two autosomal recessive variants are well described. Thirty to 50% of patients diagnosed with ARVC/D have a mutation in at least one gene. The majority of these patients have mutations in genes, which encode desmosomal proteins. Although much progress has been made in discovering some of the genetic and molecular mechanisms responsible for ARVC/D, evidence for genotype-to-phenotype correlation is still emerging. Due to incomplete penetrance and variable expressivity, the phenotypes are heterogeneous and difficult to predict. This has limited the use of gene identification as an adjunct to clinical management of affected patients.

EPIDEMIOLOGY

The estimated prevalence of ARVC/D is between 1 in 2000 and 1 in 5000, with a male predominance of nearly a 1.3:1 ratio [2,3]. ARVC/D is a cause of sudden cardiac death in people less than 35 years of age and can account for up to 12.5–24% of these cases [4,5].

Two particular subpopulations deserve mention due to founder effect including mutations in genes for transmembrane protein 43 (*TMEM43*) and phospholamban (*PLN*). In Newfoundland, Canada, identical *TMEM43* deletions were characterized in 15 families, leading to 257 individuals with a mutation [6]. In the Dutch population, mutations in *PLN* are of particular interest as a single variant *PLN* mutation has been identified in 10–15% of patients diagnosed with either dilated cardiomyopathy or ARVC/D [7].

GENETIC DETERMINANTS

Desmosomal Genes

Cardiac structural and functional integrity is accomplished by desmosomes, adherens junctions, and gap junctions located in the intercalated disks of cell-to-cell junctions [8]. ARVC/D is a disease of the cell-to-cell junctions with the majority of cases showing a predilection for mutations directly affecting desmosomal proteins. Plakophilin-2 (PKP2), plakoglobin (JUP), desmoplakin (DSP), desmoglein-2 (DSG2), and desmocollin-2 (DSC2) are the five desmosomal proteins in which mutations are known to cause ARVC/D. The desmosome is a symmetrical protein complex (Fig. 6.1) with each end residing in the cytoplasm of adjacent cells that anchors intermediate filaments to the cell surface. Aside from structural support, desmosomes have important roles in regulation of cell-to-cell signaling and can influence transcriptional regulation of genes engaged in proliferation and differentiation [8]. *PKP2*, *DSG2*, and *DSC2* mutations are all transmitted in autosomal dominant fashion. *JUP* and *DSP* can be transmitted in an autosomal recessive or dominant fashion and both may manifest as cardiocutaneous syndromes.

Nondesmosomal Genes

As previously mentioned, several nondesmosomal genes have also been implicated as disease causing: transforming growth factor beta-3 (*TGFB3*) [9], cardiac ryanodine receptor (*RYR2*) [10], *TMEM43* [6], tumor protein p63 (*TP63*) [11], desmin (*DES*) [12,13], lamin A/C (*LMNA*) [14], alpha T-catenin (*CTNNA3*) [15,16], titin (*TTN*) [17], and *PLN* [7]. For a variety of reasons, including overlap with other diseases and insufficient data,

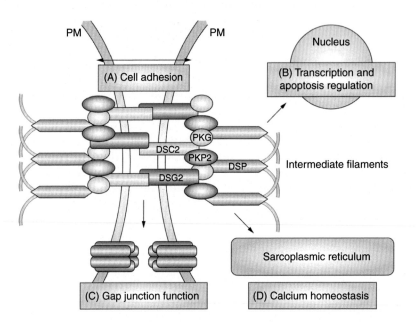

FIGURE 6.1 **The cardiac desmosome and proposed roles of the desmosome.** In (A) supporting structural stability through cell–cell adhesion, (B) regulating transcription of genes involved in adipogenesis and apoptosis, and maintaining proper electrical conductivity through regulation of (C) gap junctions and (D) calcium homeostasis. DSC2, desmocollin-2; DSG2, desmoglein-2; DSP, desmoplakin; PKG, plakoglobin; PKP2, plakophilin-2; PM, plasma membrane. *Recreated with permission from Nature Publishing Group [8].*

the pathogenetic correlation to disease of many of the nondesmosomal genes is mostly speculative at this point.

Phenotype

The phenotypes of ARVC/D are characterized as having three general clinical patterns: *classic right,* present in 39% of cases; *left dominant,* present in 5% of cases; and *biventricular,* present in 56% of cases [18]. The average age of onset is often generally in the early 30s with one study quoting a median age of onset at 26 [18–20]. It is rare to demonstrate clinical signs or symptoms prior to the age of 12 or develop onset of symptoms after the age of 60 [18,20]. Palpitations and syncope are the most common presenting symptoms [18,20].

Classic ARVC/D is a right heart dominant disease characterized by four distinct phases [21]. In the early "concealed phase," individuals are often asymptomatic with little or no RV abnormality, but still have a risk of sudden cardiac death [22]. In the "overt electrical phase," symptomatic ventricular arrhythmias are present including premature ventricular complexes, nonsustained ventricular tachycardia (VT), sustained VT, or ventricular fibrillation. At this stage, RV morphological abnormalities are present by imaging. Disease progression leads to a third phase of RV failure, where fibro-fatty replacement of normal RV myocardium leads to impaired RV dynamics and varying degrees of RV failure. Finally the disease can progress to biventricular failure, an advanced stage where fibro-fatty replacement of myocardium involves the septum and LV leading to left-sided heart failure. This stage can resemble dilated cardiomyopathy.

The left dominant pattern can be distinguished from classic ARVC/D in that it has mild right heart disease with early and prominent left ventricular disease [21]. The biventricular pattern differs from the classic pattern in that it demonstrates parallel right and left ventricular involvement rather than sequential right and left ventricular involvement. The prevalence of these subtypes may be underrepresented due to the clinical syndrome being attributed to other diseases and the lack of specific diagnostic guidelines for these entities.

Phenotypic heterogeneity of ARVC/D has suggested that environmental influences play a role in expression. Environmental influences include exercise and others are still likely undiscovered. One study showed that ARVC/D patients who participated in competitive athletics had a fivefold increased risk of sudden death compared with those who were nonathletes [23]. Due to this association, and the reports that exercise causes earlier onset of the disease, it has been recommended that patients with ARVC/D do not participate in competitive sports [24–26]. In a study of 87 individual carriers of desmosomal mutations linked to ARVC/D, it was found that participation in endurance athletics was associated with an increased risk of VT/fibrillation, congestive heart failure, and ARVC/D by revised 2010 Task Force Criteria (RTFC) [24]. Not only does this highlight the importance of exercise counseling in genetic carriers who meet RTFC for ARVC/D, but also appears to be applicable to carriers who do not meet RTFC for ARVC/D.

GENOTYPE–PHENOTYPE CORRELATIONS

Currently, knowledge of the genotype-to-phenotype correlation is progressing as more insights are made into molecular and genetic mechanisms of ARVC/D. Some studies have suggested those who are desmosomal mutation carriers have a greater incidence of VT [19] and major ventricular arrhythmic events, defined as sudden cardiac death, VT, or appropriate implantable cardioverter defibrillator shock [27], compared to nondesmosomal carriers. It has also been found that those who have genetic mutations (almost all desmosomal) for ARVC/D had a higher incidence of T-wave inversions in V1–V3 and minor structural abnormalities by RTFC [19]. Patients with ARVC/D who have more than one mutation with compound or digenic heterozygosity are common, with one study reporting that this was present in up to 16% of desmosomal carriers [28]. The same study suggests that multiple gene mutations and male sex were independent predictors of lifetime arrhythmic events [28]. Currently we have the most knowledge about the genotype–phenotype correlation of desmosomal mutation carriers since they account for the majority of ARVC/D cases. Later in this section we will discuss selected studies on the genotype–phenotype relation of *JUP*, *DSP*, and *PKP2* in more detail.

Not much is known about the molecular pathogenesis and genotype–phenotype correlation because of the relative paucity of patients and/or overlap with other syndromes. Mutations in *LMNA* cause a variety of clinical findings that include dilated cardiomyopathy and severe skeletal muscle involvement [29]. More commonly these patients have atrial arrhythmias and AV block with a high incidence of sudden cardiac death (SCD) later in the disease as CHF progresses [30]. Founder effects have allowed for better characterization of *TMEM43* and *PLN* which are both discussed later in this section. Data on *TTN* mutations have also demonstrated genotype–phenotype correlation and will also be discussed.

Plakoglobin

The involvement of the plakoglobin gene (*JUP*) in ARVC/D was discovered in patients from the Greek island of Naxos after investigation of an autosomal recessive cardiocutaneous syndrome called "Naxos Disease" involving ARVC/D, nonepidermolytic palmoplantar keratoderma, and woolly hair [31]. Cutaneous manifestations are present as early as infancy and precede cardiac manifestations [32]. Electrical manifestations such as frequent premature ventricular complexes precede structural cardiac abnormalities [32]. In this variant, the RV is always involved and progressively becomes dilated and hypokinetic with later involvement of the LV [32]. By the second decade 10% of patients have LV involvement and by the fifth decade 50% of patients have LV involvement [33]. A study of *JUP* knockout mice showed that endurance exercise accelerated RV dysfunction and arrhythmias [34]. Interestingly, ARVC/D was described in a German family occurring with an autosomal dominant mutation of *JUP* that did not have any cutaneous manifestations [35].

Desmoplakin

Carvajal syndrome, another autosomal recessive cardiocutaneous syndrome described in families from India and Ecuador, has been linked to mutations in *DSP* [36]. These families also had woolly hair and palmoplantar keratoderma, but showed predominately left-sided dilated cardiomyopathy [36]. Since the discovery of this gene, autosomal recessive and autosomal dominant mutations in *DSP* have been identified in patients with similar phenotype [8]. Some mutations in *DSP* have been found that cause isolated cardiac abnormalities without cutaneous manifestations as well as others that cause isolated cutaneous manifestations without cardiomyopathy [8].

Plakophilin-2

The most common genetic mutation found in ARVC/D is the gene encoding PKP-2, which is transmitted in an autosomal dominant fashion. Two studies have found a prevalence of mutations in *PKP2* of approximately 44% among patients with ARVC/D [19,37]. Typically carriers of the *PKP2* mutation follow the classic pattern of clinical expression, but this finding is likely due to the fact that the majority of ARVC/D patients with a genetic mutation have *PKP2*. In a study of 90 Chinese subjects with ARVC/D, the patients who had *PKP2* mutations had a greater incidence of VT (95% vs. 62%), more frequent left bundle branch type VT (80% vs. 56%), negative T-waves in V1–V3 (65% vs. 40%), and a higher proportion of induced fast VT $\geq$ 200 BPM at electrophysiology study (91% vs. 67%) when compared with the subjects who had either no identified gene or a non-*PKP2* mutation [19]. Conflicting evidence exists regarding phenotypic expression of truncating *PKP2* mutations. In a Japanese cohort, it was found that the patients with truncating *PKP2* mutations developed the disease at a younger age than those with other desmosomal mutations [38]. Another study in a Spanish cohort suggested that truncating *PKP2* mutations were associated with a later age of onset of the disease when compared to other desmosomal mutations [39].

It has been reported that *PKP2* mutations have been found at rates of 1 in 200 healthy Finnish people and as many as 6% healthy persons of non-Caucasian descent from the Netherlands [40,41]. These rates are more frequent than what would be expected in a truly abnormal desmosomal gene predicted to be disease causing. Due to a high proportion of mutations in healthy individuals together with the frequency of compound and digenic heterozygosity of mutations, it has been suggested that *PKP2* mutations may require a second mutation in *PKP2* or another desmosomal gene to cause ARVC/D [42]. Thus, family members who do not carry a mutation in *PKP2* or affected persons with known *PKP2* mutations may still be at risk of having a disease causing mutation in an unknown gene.

Titin

TTN is the largest gene expressed in mammals. Its role in cardiomyopathy has greatly expanded in recent years. It has been implicated as

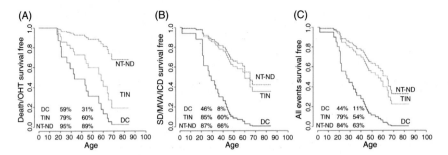

FIGURE 6.2 **Comparison of long-term natural history between desmosomal genes (PKP2, DSP, DSG2), TTN rare variant carriers, and nontitin and nondesmosomal carriers (NT-ND).** Follow-up from birth to endpoint/last follow-up evaluation. (A) Survival free from D/OHT. (B). Survival free from MVA (SD, S-VT, or appropriate discharge of ICD). (C) Survival free from the combined endpoint A + B. Percentages in the figures are referred to the survival rate at the age 30 and 50 years, respectively. ICD, implantable cardioverter defibrillator discharge; MVA, malignant ventricular arrhythmia; S-VT, sustained ventricular tachycardia; D/OHT, death or heart transplant. *Recreated with permission from BMJ Publishing Group Ltd. [27].*

the most common disease gene in up to 25% of patients with dilated cardiomyopathy. TTN has also been implicated in ARVC/D [17,43]. TTN is a nondesmosomal protein involved in cellular mechanics and it is suggested that proteolysis of a structurally weakened TTN may lead to RV dysfunction and apoptosis characteristic of ARVC/D [27]. One study compared *TTN* mutation carriers to desmosomal mutation carriers and to noncarriers. In that study, both of the desmosomal mutation carriers and *TTN* carriers had a significant increase in lifetime risk of death or heart transplant and the occurrence of malignant ventricular arrhythmias (MVAs) when compared to the noncarrier group (Fig. 6.2) [27]. Clinically, *TTN* carriers had a higher incidence of exertional dyspnea and heart failure, supraventricular arrhythmias, and bradyarrhythmias than noncarriers, but the incidence of the aforementioned clinical parameters was similar to the desmosomal carrier population [27]. Overall desmosomal carriers and *TTN* carriers also had a higher incidence of heart transplantation compared to noncarriers [27]. Echocardiographically *TTN* carriers also had more left atrial enlargement, mitral regurgitation, and RV dilation compared with desmosomal carriers and the noncarrier group [27].

TMEM43

TMEM43 is a nondesmosomal protein that is linked to families with an ARVC/D phenotype in a genetically isolated population of Newfoundland, Canada. Little is known about how mutations in *TMEM43* can lead to ARVC/D, but it is postulated that the gene contains a response element for an adipogenic transcription factor that may correlate with the fibro-fatty infiltration seen in ARVC/D [6]. In a study of 15 families with ARVC/D from this genetically isolated population, all clinically affected subjects had identical mutations in *TMEM43* [6]. Characteristics of mutation carriers without symptoms who were referred on the basis of family history that showed 57% had electrocardiographic signs of ARVC/D or an enlarged LV [6]. In this population, affected males were three times more likely to develop heart failure. Fourteen of 89 affected males developed heart failure at a median age of 63 and 7 of 83 females developed heart failure at a median age of 73 [6]. The median survival for affected subjects was significantly reduced for both males and females when compared to controls [6]. For affected males, median survival was 41 years compared to 83 years for controls and for affected females the median survival was 71 years compared to 83 for controls [6]. Primary prevention implantable cardiac defibrillator (ICD) has been studied in this group since this genotype has a particularly lethal phenotype in affected males. The current recommendation is that *TMEM43* affected males receive a primary prevention ICD [44].

Phospholamban

PLN is another nondesmosomal protein that has been implicated in ARVC/D and dilated cardiomyopathy. Like *TMEM43* it too was discovered by founder effects in the Netherlands, where as many as 10–15% of patients diagnosed with DCM or ARVC/D harbor this mutation [7]. PLN is a transmembrane sarcoplasmic reticulum protein and is a key regulator of calcium homeostasis but its mechanistic role in cardiomyopathy is not well understood [7]. In a study of 295 Dutch individuals with identical *PLN* mutations, an analysis was made comparing symptomatic carriers, in which the first referral to the cardiologist was instigated by cardiac symptoms, and asymptomatic carriers who were referred due to a family history of SCD, cardiomyopathy, and/or *PLN* mutation. The initial clinical manifestation of the symptomatic carrier group was VT or resuscitated SCD in 20%, unexplained syncope in 13%, and NYHA class III or IV heart failure in 16% [7]. Symptomatic carriers also tended to be older men, with a mean age of 48 compared to 41, in the asymptomatic carrier group [7]. Interestingly, there was no significant difference between the number of symptomatic and asymptomatic patients who met RTFC for ARVC/D, although more of the symptomatic patients had RV enlargement [7]. During follow-up, 19% of both symptomatic and asymptomatic individuals had malignant ventricular arrhythmias at a mean age of 46. Transplant or death due to heart failure occurred in 11% of these individuals at a mean age of 53

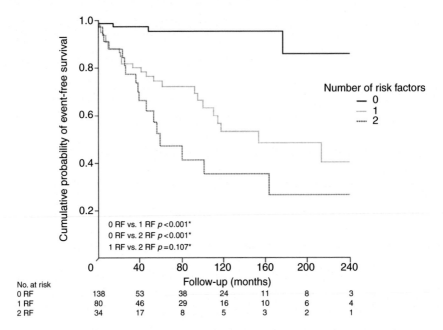

FIGURE 6.3 **Kaplan–Meier event-free survival: stratified by two independent risk factors.** The two risk factors are: left ventricular ejection fraction (LVEF) <45% and occurrence of sustained or nonsustained VT. The analysis of cardiac data from 295 PLN mutation carriers enabled the authors to perform risk stratification for MVAs (cardiopulmonary resuscitation, appropriate implantable cardioverter defibrillator intervention, or sudden cardiac death), which is essential for preventing sudden cardiac death. Independent risk factors for MVAs were LVEF <45% and sustained or nonsustained VT with hazard ratios of 4.0 (95% confidence interval [CI], 1.9–8.1) and 2.6 (95% CI, 1.5–4.5), respectively. RF indicates risk factor, the asterisks indicate the corresponding statistical significance. *Recreated with permission from Wolters Kluwer Health [7].*

[7]. In this cohort, having LVEF $\leq$ 45% and/or NSVT significantly reduced event-free survival in all mutation carriers (Fig. 6.3) [7]. Congestive heart failure, low-voltage ECG, and MVAs defined as cardiopulmonary resuscitation, appropriate implantable cardioverter defibrillator intervention, or sudden cardiac death was common among carriers [7].

FAMILY SCREENING

Diagnosis of probands relies on the RTFC in which a definite diagnosis of ARVC/D can be made by the presence of two major criteria; one major criterion and two minor criteria or four minor criteria [45]. Due to the broad phenotypic variation and autosomal dominant characteristic of the disease, Hamid et al. investigated a group of first- and second-degree

TABLE 6.1   Proposed Modification of the Original Task Force Criteria for the Diagnosis of Familial ARVC/D [43]

ARVC/D in first-degree relative plus one of the following	
1. ECG	T-wave inversion in right precordial leads (V2 and V3)
2. SAECG	Late potentials seen on signal-averaged ECG
3. Arrhythmia	LBBB-type VT on ECG, Holter monitoring, or during exercise testing
	Extrasystoles >200 over a 24-h period*
4. Structural or functional abnormality of the RV	Mild global RV dilatation and/or EF reduction with normal LV
	Mild segmental dilatation of the RV
	Regional RV hypokinesia

*>1000/24-h period in original task force criteria.
ARVC/D, arrhythmogenic right ventricular cardiomyopathy/dysplasia; ECG, electrocardiogram; EF, ejection fraction; LBBB, left bundle branch block; RV, right ventricle; SAECG, signal-averaged electrocardiography; VT, ventricular tachycardia.

family members of ARVC/D probands as adjudicated by the original Task Force criteria (TFC) [46]. They found that 28% of their cohort met TFC criteria for ARVC/D, but another 11% of the cohort had minor cardiac abnormalities [46]. They proposed alternate criteria which take into account a higher probability of first- and second-degree relatives having ARVC/D than in the general population and indicated that these criteria had increased sensitivity in detecting disease in this population (Table 6.1). Currently, the use of Hamid et al. criteria is still debated and under evaluation.

Genetic testing is currently recommended for all index cases of ARVC/D since a positive test can help guide genetic counseling, screening, and follow-up in family members. As mentioned previously, some studies have also demonstrated genotype–phenotype correlations that could have potential for risk stratification and management decisions. Identification of genetic carriers can aid in the diagnosis of ARVC/D in probands and family members together with counseling regarding physical activity for those who are carriers but do not demonstrate clinical signs or symptoms of ARVC/D.

The frequency and intervals for clinical screening of asymptomatic carriers and asymptomatic at-risk relatives, in families where no mutation has been identified, has been suggested in the Guidelines of Hershberger et al. [47]. For asymptomatic ARVC/D gene carriers, the recommendations are to have clinical screening every year between the ages 10 and 50, or at any time that signs or symptoms appear [47]. The recommended clinical screening consists of detailed clinical and family history, physical examination, electrocardiogram, echocardiogram, signal averaged ECG,

Holter monitoring, and may include an MRI [47]. If genetic testing of the proband is negative, the suggested interval for asymptomatic at-risk relatives is 3 to 5 years after the age of 10 years [47]. In a report by te Riele et al., those who carried mutations (99% desmosomal carriers) but did not have ARVC/D were unlikely to have structural progression over the 4-year follow-up period. However, one-third of this at-risk population developed some form of electrical progression [48]. These findings suggest that follow-up should be focused on electrical abnormalities with less focus on screening using imaging modalities. No standardized screening protocols exist beyond how to evaluate at-risk family members at the first visit. Based on current information it is reasonable to screen family members with a focus on electrical progression in the second decade of life with less frequent screening after the fourth decade. It is not known when screening should be stopped.

CONCLUSIONS AND FUTURE DIRECTIONS

ARVC/D is a complex disease with a heterogeneous genetic etiology. Significant advances have been made in the understanding of the genetic basis of ARVC/D. Current literature suggests that approximately 50% of patients who have ARVC/D have a desmosomal mutation, and the most common mutation is *PKP2* [19,37]. From studies of unaffected mutation carriers cardiac electrical signs and symptoms of disease precede cardiac structural abnormalities [48]. Genetic testing of those affected with ARVC/D is important as this has implications not only for the patient but also for family members. Some studies have suggested that those with genetic mutations may have a more pronounced phenotype than those without genetic mutations [19,27]. Those who have autosomal recessive forms of the disease also have a more penetrant form of ARVC/D and have cutaneous abnormalities that precede cardiac manifestations [31,36]. It is now recognized that ARVC/D is often associated with multiple potential disease causing mutations. Disease carriers with digenic or compound heterozygosity have been described as having worse a prognosis than those with single mutations, with one study showing an increased risk of arrhythmic events [28]. A relationship between *DSP* gene mutations and left ventricular involvement has been well established. However, as genetic testing becomes more widespread links between other ARVC/D causing genes and left ventricular involvement are emerging [49]. Next-generation sequencing will increase information of genotype–phenotype correlations. Future research will focus on expanding genotype–phenotype studies and integrate phenome–genome data in order to provide more accurate information on diagnosis, prognosis, risk stratification, and potential molecular therapies for affected subjects.

References

[1] Basso C, Bauce B, Corrado D, Thiene G. Pathophysiology of arrhythmogenic cardiomy-opathy. Nat Rev Cardiol 2012;9:223–33.
[2] Corrado D, Thiene G. Arrhythmogenic right ventricular cardiomyopathy/dysplasia: clinical impact of molecular genetic studies. Circulation 2006;113:1634–7.
[3] Marcus FI, Zareba W, Calkins H, Towbin JA, Basso C, Bluemke DA, Estes NA III, Picard MH, Sanborn D, Thiene G, Wichter T, Cannom D, Wilber DJ, Scheinman M, Duff H, Daubert J, Talajic M, Krahn A, Sweeney M, Garan H, Sakaguchi S, Lerman BB, Kerr C, Kron J, Steinberg JS, Sherrill D, Gear K, Brown M, Severski P, Polonsky S, McNitt S. Arrhythmogenic right ventricular cardiomyopathy/dysplasia clinical presentation and diagnostic evaluation: results from the North American Multidisciplinary Study. Heart Rhythm 2009;6:984–92.
[4] Puranik R, Gray B, Lackey H, Yeates L, Parker G, Duflou J, Semsarian C. Comparison of conventional autopsy and magnetic resonance imaging in determining the cause of sudden death in the young. J Cardiovasc Magn Reson 2014;16:44.
[5] Basso C, Corrado D, Thiene G. Cardiovascular causes of sudden death in young indi-viduals including athletes. Cardiol Rev 1999;7:127–35.
[6] Merner ND, Hodgkinson KA, Haywood AF, Connors S, French VM, Drenckhahn JD, Kupprion C, Ramadanova K, Thierfelder L, McKenna W, Gallagher B, Morris-Larkin L, Bassett AS, Parfrey PS, Young TL. Arrhythmogenic right ventricular cardiomyopathy type 5 is a fully penetrant, lethal arrhythmic disorder caused by a missense mutation in the TMEM43 gene. Am J Hum Genet 2008;82:809–21.
[7] van Rijsingen IA, van der Zwaag PA, Groeneweg JA, Nannenberg EA, Jongbloed JD, Zwinderman AH, Pinto YM, Dit Deprez RH, Post JG, Tan HL, de Boer RA, Hauer RN, Christiaans I, van den Berg MP, van Tintelen JP, Wilde AA. Outcome in phospholam-ban r14del carriers: results of a large multicentre cohort study. Circ Cardiovasc Genet 2014;7:455–65.
[8] Awad MM, Calkins H, Judge DP. Mechanisms of disease: molecular genetics of arrhyth-mogenic right ventricular dysplasia/cardiomyopathy. Nat Clin Pract Cardiovasc Med 2008;5:258–67.
[9] Beffagna G, Occhi G, Nava A, Vitiello L, Ditadi A, Basso C, Bauce B, Carraro G, Thiene G, Towbin JA, Danieli GA, Rampazzo A. Regulatory mutations in transforming growth factor-beta3 gene cause arrhythmogenic right ventricular cardiomyopathy type 1. Cardiovasc Res 2005;65:366–73.
[10] Tiso N, Stephan DA, Nava A, Bagattin A, Devaney JM, Stanchi F, Larderet G, Brahmb-hatt B, Brown K, Bauce B, Muriago M, Basso C, Thiene G, Danieli GA, Rampazzo A. Identification of mutations in the cardiac ryanodine receptor gene in families affected with arrhythmogenic right ventricular cardiomyopathy type 2 (ARVD2). Hum Mol Ge-netV 10 2001;189–94.
[11] Valenzise M, Arrigo T, De Luca F, Privitera A, Frigiola A, Carando A, Garelli E, Silengo M. R298Q mutation of p63 gene in autosomal dominant ectodermal dysplasia associated with arrhythmogenic right ventricular cardiomyopathy. Eur J Med Genet 2008;51:497–500.
[12] Lorenzon A, Beffagna G, Bauce B, De Bortoli M, Li Mura IE, Calore M, Dazzo E, Basso C, Nava A, Thiene G, Rampazzo A. Desmin mutations and arrhythmogenic right ventricu-lar cardiomyopathy. Am J Cardiol 2013;111:400–5.
[13] Otten E, Asimaki A, Maass A, van Langen IM, van der Wal A, de Jonge N, van den Berg MP, Saffitz JE, Wilde AA, Jongbloed JD, van Tintelen JP. Desmin mutations as a cause of right ventricular heart failure affect the intercalated disks. Heart Rhythm 2010;7:1058–64.
[14] Quarta G, Syrris P, Ashworth M, Jenkins S, Zuborne Alapi K, Morgan J, Muir A, Pantazis A, McKenna WJ, Elliott PM. Mutations in the Lamin A/C gene mimic arrhythmogenic right ventricular cardiomyopathy. Eur Heart J 2012;33:1128–36.

[15] Gandjbakhch E, Vite A, Gary F, Fressart V, Donal E, Simon F, Hidden-Lucet F, Komajda M, Charron P, Villard E. Screening of genes encoding junctional candidates in arrhythmogenic right ventricular cardiomyopathy/dysplasia. Europace V 15 2013;1522–5.

[16] van Hengel J, Calore M, Bauce B, Dazzo E, Mazzotti E, De Bortoli M, Lorenzon A, Li Mura IE, Beffagna G, Rigato I, Vleeschouwers M, Tyberghein K, Hulpiau P, van Hamme E, Zaglia T, Corrado D, Basso C, Thiene G, Daliento L, Nava A, van Roy F, Rampazzo A. Mutations in the area composita protein alphaT-catenin are associated with arrhythmogenic right ventricular cardiomyopathy. Eur Heart J 2013;34:201–10.

[17] Taylor M, Graw S, Sinagra G, Barnes C, Slavov D, Brun F, Pinamonti B, Salcedo EE, Sauer W, Pyxaras S, Anderson B, Simon B, Bogomolovas J, Labeit S, Granzier H, Mestroni L. Genetic variation in titin in arrhythmogenic right ventricular cardiomyopathy-overlap syndromes. Circulation 2011;124:876–85.

[18] Sen-Chowdhry S, Syrris P, Ward D, Asimaki A, Sevdalis E, McKenna WJ. Clinical and genetic characterization of families with arrhythmogenic right ventricular dysplasia/cardiomyopathy provides novel insights into patterns of disease expression. Circulation 2007;115:1710–20.

[19] Bao J, Wang J, Yao Y, Wang Y, Fan X, Sun K, He DS, Marcus FI, Zhang S, Hui R, Song L. Correlation of ventricular arrhythmias with genotype in arrhythmogenic right ventricular cardiomyopathy. Circ Cardiovasc Genet 2013;6:552–6.

[20] Dalal D, Nasir K, Bomma C, Prakasa K, Tandri H, Piccini J, Roguin A, Tichnell C, James C, Russell SD, Judge DP, Abraham T, Spevak PJ, Bluemke DA, Calkins H. Arrhythmogenic right ventricular dysplasia: a United States experience. Circulation 2005;112:3823–32.

[21] Sen-Chowdhry S, Morgan RD, Chambers JC, McKenna WJ. Arrhythmogenic cardiomyopathy: etiology, diagnosis, and treatment. Annu Rev Med 2010;61:233–53.

[22] Thiene G, Nava A, Corrado D, Rossi L, Pennelli N. Right ventricular cardiomyopathy and sudden death in young people. N Engl J Med 1988;318:129–33.

[23] Corrado D, Basso C, Rizzoli G, Schiavon M, Thiene G. Does sports activity enhance the risk of sudden death in adolescents and young adults? J Am Coll Cardiol 2003;42:1959–63.

[24] James CA, Bhonsale A, Tichnell C, Murray B, Russell SD, Tandri H, Tedford RJ, Judge DP, Calkins H. Exercise increases age-related penetrance and arrhythmic risk in arrhythmogenic right ventricular dysplasia/cardiomyopathy-associated desmosomal mutation carriers. J Am Coll Cardiol 2013;62:1290–7.

[25] Maron BJ, Ackerman MJ, Nishimura RA, Pyeritz RE, Towbin JA, Udelson JE. Task Force 4: HCM and other cardiomyopathies, mitral valve prolapse, myocarditis, and Marfan syndrome. J Am Coll Cardiol 2005;45:1340–5.

[26] Maron BJ, Chaitman BR, Ackerman MJ, Bayes de Luna A, Corrado D, Crosson JE, Deal BJ, Driscoll DJ, Estes NA 3rd, Araujo CG, Liang DH, Mitten MJ, Myerburg RJ, Pelliccia A, Thompson PD, Towbin JA, Van Camp SP. Working Groups of the American Heart Association Committee on Exercise Cardiac Rehabilitation and Prevention, Councils on Clinical Cardiology and Cardiovascular Disease in the Young. Recommendations for physical activity and recreational sports participation for young patients with genetic cardiovascular diseases. Circulation 2004;109:2807–16.

[27] Brun F, Barnes CV, Sinagra G, Slavov D, Barbati G, Zhu X, Graw SL, Spezzacatene A, Pinamonti B, Merlo M, Salcedo EE, Sauer WH, Taylor MR, Mestroni L. Familial Cardiomyopathy Registry. Titin and desmosomal genes in the natural history of arrhythmogenic right ventricular cardiomyopathy. J Med Genet 2014;51:669–76.

[28] Rigato I, Bauce B, Rampazzo A, Zorzi A, Pilichou K, Mazzotti E, Migliore F, Marra MP, Lorenzon A, De Bortoli M, Calore M, Nava A, Daliento L, Gregori D, Iliceto S, Thiene G, Basso C, Corrado D. Compound and digenic heterozygosity predicts lifetime arrhythmic outcome and sudden cardiac death in desmosomal gene-related arrhythmogenic right ventricular cardiomyopathy. Circ Cardiovasc Genet 2013;6:533–42.

[29] Taylor MR, Fain PR, Sinagra G, Robinson ML, Robertson AD, Carniel E, Di Lenarda A, Bohlmeyer TJ, Ferguson DA, Brodsky GL, Boucek MM, Lascor J, Moss AC, Li WL, Stetler GL, Muntoni F, Bristow MR, Mestroni L. Familial Dilated Cardiomyopathy Registry Research Group. Natural history of dilated cardiomyopathy due to lamin A/C gene mutations. J Am Coll Cardiol 2003;41:771–80.

[30] Fatkin D, MacRae C, Sasaki T, Wolff MR, Porcu M, Frenneaux M, Atherton J, Vidaillet HJ Jr, Spudich S, De Girolami U, Seidman JG, Seidman C, Muntoni F, Muehle G, Johnson W, McDonough B. Missense mutations in the rod domain of the lamin A/C gene as causes of dilated cardiomyopathy and conduction-system disease. N Engl J Med 1999;341:1715–24.

[31] McKoy G, Protonotarios N, Crosby A, Tsatsopoulou A, Anastasakis A, Coonar A, Norman M, Baboonian C, Jeffery S, McKenna WJ. Identification of a deletion in plakoglobin in arrhythmogenic right ventricular cardiomyopathy with palmoplantar keratoderma and woolly hair (Naxos disease). Lancet 2000;355:2119–24.

[32] Protonotarios N, Tsatsopoulou A, Anastasakis A, Sevdalis E, McKoy G, Stratos K, Gatzoulis K, Tentolouris K, Spiliopoulou C, Panagiotakos D, McKenna W, Toutouzas P. Genotype–phenotype assessment in autosomal recessive arrhythmogenic right ventricular cardiomyopathy (Naxos disease) caused by a deletion in plakoglobin. J Am Coll Cardiol 2001;38:1477–84.

[33] Tsatsopoulou AA, Protonotarios NI, McKenna WJ. Arrhythmogenic right ventricular dysplasia, a cell adhesion cardiomyopathy: insights into disease pathogenesis from preliminary genotype–phenotype assessment. Heart 2006;92:1720–3.

[34] Kirchhof P, Fabritz L, Zwiener M, Witt H, Schafers M, Zellerhoff S, Paul M, Athai T, Hiller KH, Baba HA, Breithardt G, Ruiz P, Wichter T, Levkau B. Age- and training-dependent development of arrhythmogenic right ventricular cardiomyopathy in heterozygous plakoglobin-deficient mice. Circulation 2006;114:1799–806.

[35] Asimaki A, Syrris P, Wichter T, Matthias P, Saffitz JE, McKenna WJ. A novel dominant mutation in plakoglobin causes arrhythmogenic right ventricular cardiomyopathy. Am J Hum Genet 2007;81:964–73.

[36] Norgett EE, Hatsell SJ, Carvajal-Huerta L, Cabezas JC, Common J, Purkis PE, Whittock N, Leigh IM, Stevens HP, Kelsell DP. Recessive mutation in desmoplakin disrupts desmoplakin-intermediate filament interactions and causes dilated cardiomyopathy, woolly hair and keratoderma. Hum Mol Genet 2000;9:2761–6.

[37] Dalal D, Molin LH, Piccini J, Tichnell C, James C, Bomma C, Prakasa K, Towbin JA, Marcus FI, Spevak PJ, Bluemke DA, Abraham T, Russell SD, Calkins H, Judge DP. Clinical features of arrhythmogenic right ventricular dysplasia/cardiomyopathy associated with mutations in plakophilin-2. Circulation 2006;113:1641–9.

[38] Ohno S, Nagaoka I, Fukuyama M, Kimura H, Itoh H, Makiyama T, Shimizu A, Horie M. Age-dependent clinical and genetic characteristics in Japanese patients with arrhythmogenic right ventricular cardiomyopathy/dysplasia. Circ J 2013;77:1534–42.

[39] Alcalde M, Campuzano O, Berne P, Garcia-Pavia P, Doltra A, Arbelo E, Sarquella-Brugada G, Iglesias A, Alonso-Pulpon L, Brugada J, Brugada R. Stop-gain mutations in PKP2 are associated with a later age of onset of arrhythmogenic right ventricular cardiomyopathy. PloS One 2014;9:e100560.

[40] Kapplinger JD, Landstrom AP, Salisbury BA, Callis TE, Pollevick GD, Tester DJ, Cox MG, Bhuiyan Z, Bikker H, Wiesfeld AC, Hauer RN, van Tintelen JP, Jongbloed JD, Calkins H, Judge DP, Wilde AA, Ackerman MJ. Distinguishing arrhythmogenic right ventricular cardiomyopathy/dysplasia-associated mutations from background genetic noise. J Am Coll Cardiol 2011;57:2317–27.

[41] Lahtinen AM, Lehtonen E, Marjamaa A, Kaartinen M, Helio T, Porthan K, Oikarinen L, Toivonen L, Swan H, Jula A, Peltonen L, Palotie A, Salomaa V, Kontula K. Population-prevalent desmosomal mutations predisposing to arrhythmogenic right ventricular cardiomyopathy. Heart Rhythm 2011;8:1214–21.

[42] Marcus FI, Edson S, Towbin JA. Genetics of arrhythmogenic right ventricular cardiomy-opathy: a practical guide for physicians. J Am Coll Cardiol 2013;61:1945–8.

[43] Herman DS, Lam L, Taylor MR, Wang L, Teekakirikul P, Christodoulou D, Conner L, DePalma SR, McDonough B, Sparks E, Teodorescu DL, Cirino AL, Banner NR, Pennell DJ, Graw S, Merlo M, Di Lenarda A, Sinagra G, Bos JM, Ackerman MJ, Mitchell RN, Murry CE, Lakdawala NK, Ho CY, Barton PJ, Cook SA, Mestroni L, Seidman JG, Seidman CE. Truncations of titin causing dilated cardiomyopathy. N Engl J Med 2012;366:619–28.

[44] Hodgkinson KA, Parfrey PS, Bassett AS, Kupprion C, Drenckhahn J, Norman MW, Thierfelder L, Stuckless SN, Dicks EL, McKenna WJ, Connors SP. The impact of implantable cardioverter-defibrillator therapy on survival in autosomal-dominant arrhythmogenic right ventricular cardiomyopathy (ARVD5). J Am Coll Cardiol 2005;45:400–8.

[45] Marcus FI, McKenna WJ, Sherrill D, Basso C, Bauce B, Bluemke DA, Calkins H, Corrado D, Cox MG, Daubert JP, Fontaine G, Gear K, Hauer R, Nava A, Picard MH, Protonotarios N, Saffitz JE, Sanborn DM, Steinberg JS, Tandri H, Thiene G, Towbin JA, Tsatsopoulou A, Wichter T, Zareba WAT Diagnosis of arrhythmogenic right ventricular cardiomyopathy/dysplasia: proposed modification of the task force criteria. Circulation 2010;121:1533–41.

[46] Hamid MS, Norman M, Quraishi A, Firoozi S, Thaman R, Gimeno JR, Sachdev B, Rowland E, Elliott PM, McKenna WJ. Prospective evaluation of relatives for familial arrhythmogenic right ventricular cardiomyopathy/dysplasia reveals a need to broaden diagnostic criteria. J Am Coll Cardiol 2002;40:1445–50.

[47] Hershberger RE, Lindenfeld J, Mestroni L, Seidman CE, Taylor MR, Towbin JA. Heart Failure Society of America. Genetic evaluation of cardiomyopathy – a Heart Failure Society of America practice guideline. J Card Fail 2009;15:83–97.

[48] te Riele AS, James CA, Rastegar N, Bhonsale A, Murray B, Tichnell C, Judge DP, Bluemke DA, Zimmerman SL, Kamel IR, Calkins H, Tandri H. Yield of serial evaluation in at-risk family members of patients with ARVD/C. J Am Coll Cardiol 2014;64:293–301.

[49] Haas J, Frese KS, Peil B, Kloos W, Keller A, Nietsch R, Feng Z, Muller S, Kayvanpour E, Vogel B, Sedaghat-Hamedani F, Lim WK, Zhao X, Fradkin D, Kohler D, Fischer S, Franke J, Marquart S, Barb I, Li DT, Amr A, Ehlermann P, Mereles D, Weis T, Hassel S, Kremer A, King V, Wirsz E, Isnard R, Komajda M, Serio A, Grasso M, Syrris P, Wicks E, Plagnol V, Lopes L, Gadgaard T, Eiskjaer H, Jorgensen M, Garcia-Giustiniani D, Ortiz-Genga M, Crespo-Leiro MG, Deprez RH, Christiaans I, van Rijsingen IA, Wilde AA, Waldenstrom A, Bolognesi M, Bellazzi R, Morner S, Bermejo JL, Monserrat L, Villard E, Mogensen J, Pinto YM, Charron P, Elliott P, Arbustini E, Katus HA, Meder B. Atlas of the clinical genetics of human dilated cardiomyopathy. Eur Heart J 2014;36:1123–1135a.

Diagnostic Evaluation of Children with Known or Suspected ARVC/D

Frank I. Marcus, Aiden Abidov

**Department of Medicine/Division of Cardiology
and Department of Medical Imaging,
University of Arizona, Tucson, AZ, USA**

The ARVC/D phenotype is present mostly in young to middle age adulthood. Accordingly, performing CMR in a pediatric population may not be clinically effective. This situation becomes even more complex in asymptomatic children with a positive ARVC/D genotype, which is a frequent clinical finding in families of ARVC/D patients. The individual with the diagnosis of ARVC/D and with associated genetic abnormality has a 50% chance that his or her children will have the genetic abnormality. Therefore, at what age should the child of an affected parent be evaluated for the clinical manifestations of the disease? What methods should be used to test if the child is clinically affected? Is sudden arrhythmic death a problem in young children with the genetic abnormality? Is it possible to prevent the disease from developing in the offspring or at least slow its development? There are answers for some but not all of these important questions.

Let us first address the problem of inheritability of ARVC/D. Genetic testing should be performed in a child when a parent has one of the five desmosomal genetic abnormalities of ARVC/D. The reasons are persuasive. If the child has the same genetic defect as the parent, then the child has inherited the risk of becoming clinically affected with ARVC/D. There is the unlikely possibility of the parent having two abnormal desmosomal genes that occurs in 4–17% of patients with ARVC/D [1–3]. If the child also has more than one abnormal gene, then the likelihood of the child developing the disease, as well as an increased possibility of having severe clinical

Cardiac MRI in the Diagnosis, Clinical Management and Prognosis of Arrhythmogenic Right Ventricular Cardiomyopathy/Dysplasia

symptoms, is substantial. Conversely, if the parent has an abnormal desmosomal gene but the child does not, then the child's risk is minimal.

What if the parent has the clinical disease but does not have any of the desmosomal abnormalities? If so, there is probably no reason to check the child for the desmosomal abnormality, since it is extremely unlikely that a known desmosomal gene will be found in the child. However, the child of the patient with ARVC/D who does not have a mutation still has a risk of developing symptomatic disease. In the large multicenter study by Groeneweg et al. [3], of the 385 family members with an identified genetic mutation, 61 (16%) developed symptoms of ARVC/D and 51 (13%) met TFC; of the 152 family members without mutations, 13 (8.5%) were symptomatic and 10 (6.6%) met TFC (Fig. 7.1). Those 152 family members were considered without mutations because no mutation could be identified in their respective index patients. This indicates that the child of the affected patient who does not have an abnormal gene for ARVC/D has a lesser but not insignificant risk of inheriting the disease. Under these circumstances, the child of the affected parent should also be advised not to engage in endurance athletics since there is an increased risk of developing the disease [4]. These observations are probably due to the presence of yet unknown genetic abnormalities.

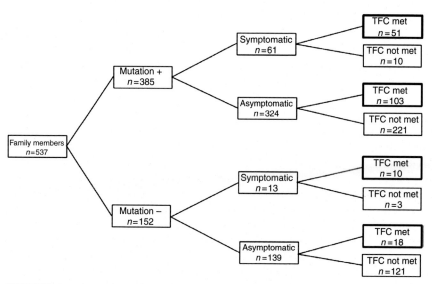

FIGURE 7.1 **Schematic representation of ARVC/D penetrance in 537 family members presenting alive.** Legend from the original paper: in 385 family members, the same mutation(s) as in their respective index-patients was identified (mutation +). The remaining 152 family members were considered "without mutations," since no mutation could be identified in their respective index-patient (mutation −). Fulfillment of TFC for ARVC/D (indicated by the thick black boxes) correlated with symptoms at first evaluation. Nonetheless, a minority of asymptomatic family members also had ARVC/D diagnosis. This underscores the importance of family screening in ARVC/D. *Reproduced with permission from Ref. [3].*

If the genetic defect is present in the child, how and when is the child likely to develop clinical manifestations including ECG abnormalities, ventricular premature beats, nonsustained or sustained ventricular tachy-cardia (VT), or have an arrhythmic death?

For the purpose of this discussion, it is important to define childhood [1]. Childhood may be classified as ≤10 years of age, from age 11 to 14, and >14 years of age. The reason for this age classification is that ARVC/D has different clinical manifestations in different age groups. For example, it is extremely rare for an arrhythmic death due to ARVC/D to occur below the age of 10 years [3,5]. Therefore, it is likely unnecessary to consider implanting an ICD in a gene-positive child before age 10 for the purpose of preventing arrhythmic death in childhood.

The earliest manifestation of the disease appears to be electrical including ventricular arrhythmias such as premature ventricular contractions (PVCs), nonsustained or sustained VT. Therefore, obtaining an ECG, Holter moni-toring, or exercise testing is a reasonable approach to see if the child has developed clinical evidence of the ARVC/D. Structural abnormalities, such as right ventricular enlargement or focal akinesia or dyskinesia, appear later than the electrical findings (see Chapter 3, Fig. 3.5), although this is contro-versial. The appearance of electrical findings prior to the detection of struc-tural abnormalities by MRI is regarded as the usual sequence of events and has been demonstrated in a mouse model of ARVC/D [6]. This is an impor-tant concept, since it impacts the decision of how to screen children for the clinical presence of the disease. Based on the above assumptions, it would appear reasonable to perform serial ECGs, Holter monitoring, and possibly exercise testing to determine whether there is electrical instability during childhood in genetically affected children. Hamid et al. [7] reported that more than 200 PVCs/24 h should be one of the determinants of whether or not a first-degree relative has ARVC/D.

A baseline MRI may also be obtained as the study does not carry a risk of radiation exposure. This testing should probably be initiated at ages 9 to 10, especially if any electrical abnormalities are present. Positive CMR findings shift the probability of ARVC/D in the young patient from probable to definite (Fig. 7.2) [8]. Tests for electrical ab-normalities should be repeated at yearly intervals or sooner if there are symptoms such as palpitations or evidence of VT. The question of how often to repeat these tests is unknown but it is suggested that they be performed at yearly intervals until age 25. If negative, then ECGs and Holter studies may be repeated every 3 years thereafter until ages 40–50; after this age, and if the patient demonstrates consistently normal results, the tests may be obtained at 5-year intervals (Fig. 7.3).

These recommendations are based on several studies that have addressed these questions. Bauce et al. [1] reviewed data on 53 subjects (31 males and 22 females) belonging to 27 families who were evaluated before age 18 and who carried desmosomal mutations. Of these 53 subjects, 26 (49%) had a genetic mutation in desmoplakin, 12 (23%) in plakophilin-2, 6 (11%)

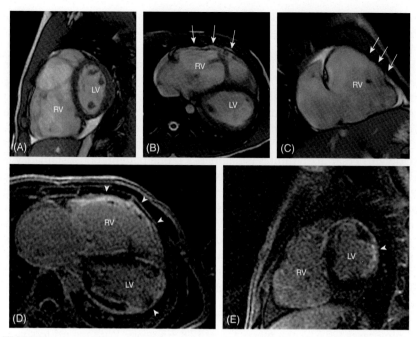

FIGURE 7.2   **CMR demonstrating evidence of the ARVC/D in a 13-year-old boy with palpitations, sustained VT requiring cardioversion, and epsilon waves on his ECG.** Short Axis (A), axial (B) and dedicated RV (C) SSFP images. Axial (D) and short axis (E) delayed hyperenhancement images. Focal RV free wall aneurysms (arrows); areas of the RV and the LV delayed hyperenhancement (arrowheads). *Reproduced with permission from Ref. [8].*

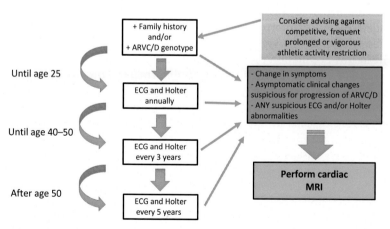

FIGURE 7.3   **Possible clinical algorithm of the follow-up for children with positive ARVC/D genotype.**

in desmoglein-2, and 9 (17%) had multiple mutations in one or two different genes. The data were divided according to age at clinical presentation as follows: Group A, ≤10 years; Group B, 11–14 years; and Group C, 14–18 years. None of the 16 subjects in Group A (mean age 7.2 ± 2.1) years fulfilled the 1994 Task Force Diagnostic Criteria. Among the 18 patients belonging to Group B (mean age 12.6 ± 1 year) there were 12 (67%) who were studied due to a family history of ARVC/D, two for chest pain, three for ventricular arrhythmias, and one because of abnormal ECG findings. Of these, six patients (33%) were diagnosed with ARVC/D. All of them had ventricular arrhythmias (sustained VT: one patient, nonsustained VT: three patients, and isolated PVCs: two patients). The mean follow-up was 9 ± 7 years. Among the 19 subjects in Group C (14–18 years), 13 (68%) were studied due to a family history of ARVC/D and the other 6 (32%) were studied for arrhythmic symptoms. ARVC/D was diagnosed in 8/19 patients (42%) of this group. All the affected patients had ventricular arrhythmias (sustained VT: four patients; nonsustained VT: two patients; and PVCs: two patients). Echocardiograms showed severe disease in two, moderate in five, and mild in one. It is important to note that the majority of these subjects (n = 39, 74%) showed no signs or symptoms of the disease. Of the 40 family members, a diagnosis of ARVC/D using the 2010 TFC was observed in nine cases (22.5%) at a mean age of 17.8 ± 5 years. With regard to structural abnormalities, none of the subjects below the age of 10 in Group A had abnormal 2D echocardiograms. At first assessment, these were found in six cases: none in Group A, six in Group B, and six in Group C. Of note, only 21 (43%) of 53 patients in this study had CMR; of those, 8 patients had an abnormal study, confirming the diagnosis of ARVC/D; of the remaining 13 patients, 12 (57%) had an abnormal CMR with evidence of delayed hyperenhancement in either the right ventricular (RV) or both ventricles. Thus, in this relatively large group of pediatric patients with detailed follow-up information, clinical ARVC/D phenotype developed in adolescence and was very uncommon in early childhood. Nevertheless, CMR in this population provided additional diagnostic information even in cases with no TFC, by demonstrating abnormal findings (delayed hyperenhancement) in the majority of the gene-positive pediatric patients.

Among these cases, a desmoplakin mutation was associated with more severe disease progression based on the right and left ventricular involvement during follow-up. The authors concluded that those with the hereditary disease should not engage in competitive sports and that CMR should be repeated yearly in those with gene-positive abnormalities. This report is important because of the long careful follow-up of a relatively large number of individuals who had a family history of ARVC/D and were stratified according to desmosomal genetic abnormalities. The authors concluded with the statement, "Management of asymptomatic gene mutation carriers remains the main clinical challenge."

This study did not show more severe disease or clinical progression based on the presence of more than one abnormal desmosomal gene in an individual. However, other studies have indicated that the presence of more than one desmosomal genetic abnormality is associated with more severe symptoms [2,9].

It is of interest that there were no arrhythmic deaths in Group A below the age of 10. This is consistent with the publication by Pilmer et al. [6] who performed a retrospective population-based study of sudden cardiac death cases of all causes in a 5-year period among individuals age 1–19 years in Ontario, Canada. Definite features of ARVC/D were found in nine cases (8%). All but one occurred in the age range of 14–18 years. One case of sudden death occurring at the age of 10.9 years in a child who had severe disease at autopsy.

Dr. Gaetano Thiene, pathologist from the University of Padua, Italy, has observed that the pathological changes of ARVC/D are seldom present before puberty (personal communication).

As previously mentioned, the role of CMR in the diagnosis of ARVC/D in children and adolescents is controversial. A recent publication by Etoom et al. [10] evaluated the CMR studies in 213 pediatric patients. A total of 23 (16%) of the patients met the modified TFC, 32 (23%) were borderline, 37 (26%) possible, and 50 (35%) had no criteria for ARVC/D. The reasons for referral included a family history in 100 (47%), cardiac symptoms in 81 (38%), ventricular arrhythmias in 75 (35%), incidental abnormal ECG findings in 4 (2%), and incidental abnormal echocardiographic findings in 6 (3%). One-third of the study population (44 subjects) had genetic testing. A pathogenic mutation was found in 4 of 9 (44%) with definite, 3 of 13 (27%) with borderline, 1 of 20 (5%) with possible, and 0 of 7 with no ARVC/D. It should be noted that right ventricular functional abnormalities including hypokinesis were considered positive although this definition was excluded from the 2010 TFC based on the opinion that hypokinesis was too subjective to be included as a criteria. Nevertheless, the authors of this chapter stated that this was utilized since most children were expected to be in the early stages of ARVC/D. This definition may have influenced the relatively high incidence of regional wall motion abnormalities reported in the study. The authors concluded that the criteria for abnormal CMR made the greatest contribution to the performance of the diagnostic score for ARVC/D in children. They also noted that there was poor agreement of CMR findings and overall wall motion score with echocardiographic findings of RV dysfunction and enlargement. Questions relating to the applicability of these reported observations have been raised in a subsequent letter to the editor [11].

The question of whether electrical abnormalities by electrocardiology or Holter monitoring precede detected abnormalities in ARVC/D mutation carriers was investigated by te Riele et al. [12]. These authors evaluated 69 patients with a mean age of 27 ± 15.3 years who had ARVC/D-associated pathogenic mutations and did not have sustained ventricular arrhythmias by ECG or 24-h Holter monitoring. There were 42 patients (61%) who had electrical abnormalities, of whom 20 (48%) had abnormal results by CMR. Only 1 of 27 patients (4%) without electrical abnormalities had abnormal CMR results. Over a mean follow-up 5.8 ± 4.4 years, 11 patients (16%) experienced sustained ventricular arrhythmias and these occurred exclusively in patients with both electrical abnormalities and abnormal CMR results.

The majority of studies indicate that electrical abnormalities on ECG or Holter monitoring are more prevalent than structural changes by CMR. Additional information to substantiate this conclusion was reported in patients with homozygous disease who had Naxos disease [13].

In a recent report that compared the clinical characteristics and outcome of patients with the clinical onset of ARVC at a mean of 15.3 ± 2.4 years ($n = 41$) with adults aged 38.6 ± 13.4 years ($n = 223$) the major difference was that there was a higher incidence of presentation with sudden cardiac death or resuscitated sudden cardiac death in the younger age group $n = 19$ (25%) as compared with 39 (9%) of similar in the older age group ($p = 0.01$). In contrast, the older age patients were more likely to present with sustained monomorphic VT $n = 151$ (35%) vs 16 (21%; $p = 0.017$) [14].

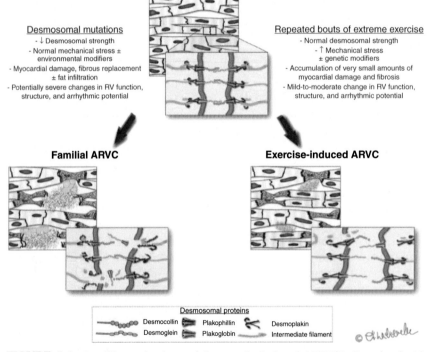

FIGURE 7.4 **Possible mechanisms of the exercise-induced ARVC/D.** *Reproduced with permission from Ref. [15].*

In conclusion, it is reasonable to suggest that ECGs and Holter monitoring be performed at yearly intervals after the age of 8 or 10 years, supplemented by MRI studies if electrical abnormalities are present. Evidence has accumulated both experimentally and clinically that frequent, prolonged, and vigorous athletic activity is a risk factor for sudden cardiac death and can exacerbate the rate of progression of the disease [4,15] (Fig. 7.4). Therefore, these types of activity should be avoided by probands or by family members who are prone to develop the disease [16]. Therapy by antiarrhythmic drugs, insertion of an ICD alone or in combination in asymptomatic children of probands with ARVC/D will require further evaluation.

References

[1] Bauce B, Rampazzo A, Basso C, Mazzotti E, Rigato I, Steriotis A, Beffagna G, Lorenzon A, De Bortoli M, Pilichou K, Marra MP, Corbetti F, Daliento L, Iliceto S, Corrado D, Thiene G, Nava A. Clinical phenotype and diagnosis of arrhythmogenic right ventricular cardiomyopathy in pediatric patients carrying desmosomal gene mutations. Heart Rhythm 2011;8:1686–95.

[2] Xu T, Yang Z, Vatta M, Rampazzo A, Beffagna G, Pilichou K, Scherer SE, Saffitz J, Kravitz J, Zareba W, Danieli GA, Lorenzon A, Nava A, Bauce B, Thiene G, Basso C, Calkins H, Gear K, Marcus F, Towbin JA. Compound and digenic heterozygosity contributes to arrhythmogenic right ventricular cardiomyopathy. J Am Coll Cardiol 2010;55:587–97.

[3] Groeneweg JA, Bhonsale A, James CA, Te Riele AS, Dooijes D, Tichnell C, Murray B, Wiesfeld AC, Sawant AC, Kassamali B, Atsma DE, Volders PG, de Groot NM, de Boer K, Zimmerman SL, Kamel IR, van der Heijden JF, Russell SD, Cramer MJ, Tedford RJ, Doevendans PA, van Veen TA, Tandri H, Wilde AA, Judge DP, van Tintelen JP, Hauer RN, Calkins H. Clinical presentation, long-term follow-up, and outcomes of 1001 arrhythmogenic right ventricular dysplasia/cardiomyopathy patients and family members. Circ Cardiovasc Genet 2015;8:437–46.

[4] Sawant AC, Bhonsale A, te Riele AS, Tichnell C, Murray B, Russell SD, Tandri H, Tedford RJ, Judge DP, Calkins H, James CA. Exercise has a disproportionate role in the pathogenesis of arrhythmogenic right ventricular dysplasia/cardiomyopathy in patients without desmosomal mutations. J Am Heart Assoc 2014;3:e001471.

[5] Pilmer CM, Kirsh JA, Hildebrandt D, Krahn AD, Gow RM. Sudden cardiac death in children and adolescents between 1 and 19 years of age. Heart Rhythm 2014;11:239–45.

[6] Gomes J, Finlay M, Ahmed AK, Ciaccio EJ, Asimaki A, Saffitz JE, Quarta G, Nobles M, Syrris P, Chaubey S, McKenna WJ, Tinker A, Lambiase PD. Electrophysiological abnormalities precede overt structural changes in arrhythmogenic right ventricular cardiomyopathy due to mutations in desmoplakin-a combined murine and human study. Eur Heart J 2012;33:1942–53.

[7] Hamid MS, Norman M, Quraishi A, Firoozi S, Thaman R, Gimeno JR, Sachdev B, Rowland E, Elliott PM, McKenna WJ. Prospective evaluation of relatives for familial arrhythmogenic right ventricular cardiomyopathy/dysplasia reveals a need to broaden diagnostic criteria. J Am Coll Cardiol 2002;40:1445–50.

[8] Saprungruang A, Tumkosit M, Kongphatthanayothin A. The presence of epsilon waves in all precordial leads (v1 -v6) in a 13-year-old boy with arrhythmogenic right ventricular dysplasia (ARVD). Ann Noninvasive Electrocardiol 2013;18:484–6.

[9] Rigato I, Bauce B, Rampazzo A, Zorzi A, Pilichou K, Mazzotti E, Migliore F, Marra MP, Lorenzon A, De Bortoli M, Calore M, Nava A, Daliento L, Gregori D, Iliceto S, Thiene G, Basso C, Corrado D. Compound and digenic heterozygosity predicts lifetime arrhythmic

outcome and sudden cardiac death in desmosomal gene-related arrhythmogenic right ventricular cardiomyopathy. Circ Cardiovasc Genet 2013;6:533–42.

[10] Etoom Y, Govindapillai S, Hamilton R, Manlhiot C, Yoo SJ, Farhan M, Sarikouch S, Peters B, McCrindle BW, Grosse-Wortmann L. Importance of CMR within the task force criteria for the diagnosis of ARVC in children and adolescents. J Am Coll Cardiol 2015;65: 987–95.

[11] te Riele AS, Marcus FI, James CA, et al. The value of cardiac magnetic resonance imaging in evaluation of pediatric patients for arrhythmogenic right ventricular dysplasia/cardiomyopathy. J Am Coll Cardiol 2015;66:873–4.

[12] te Riele AS, Bhonsale A, James CA, Rastegar N, Murray B, Burt JR, Tichnell C, Madhavan S, Judge DP, Bluemke DA, Zimmerman SL, Kamel IR, Calkins H, Tandri H. Incremental value of cardiac magnetic resonance imaging in arrhythmic risk stratification of arrhythmogenic right ventricular dysplasia/cardiomyopathy-associated desmosomal mutation carriers. J Am Coll Cardiol 2013;62:1761–9.

[13] Protonotarios N, Anastasakis A, Antoniades L, Chlouverakis G, Syrris P, Basso C, Asimaki A, Theopistou A, Stefanadis C, Thiene G, McKenna WJ, Tsatsopoulou A. Arrhythmogenic right ventricular cardiomyopathy/dysplasia on the basis of the revised diagnostic criteria in affected families with desmosomal mutations. Eur Heart J 2011;32:1097–104.

[14] te Riele ASJM, James CA, Sawant AC, et al. Arrhythmogenic right ventricular dysplasia/cardiomyopathy in the pediatric population. J Am Coll Cardiol CEP 2015;1:551–60.

[15] Heidbuchel H, Prior DL, La Gerche A. Ventricular arrhythmias associated with long-term endurance sports: what is the evidence? Br J Sports Med. 2012;46:Suppl 1:i44–50.

[16] Thiene G. The research venture in arrhythmogenic right ventricular cardiomyopathy: a paradigm of translational medicine. Eur Heart J 2015;36:837–46.

Differential Diagnosis of ARVC/D

Aiden Abidov, Frank I. Marcus

Department of Medicine/Division of Cardiology
and Department of Medical Imaging,
University of Arizona, Tucson, AZ, USA

RV SIZE AND FUNCTION-BASED ARVC/D CRITERIA: NORMAL VARIANTS AND ARTIFACTS

The Task Force criteria (TFC) [1] are focused on the morphofunctional changes in the right ventricle (RV). They are categorized as either major or minor diagnostic markers of arrhythmogenic right ventricular cardiomyopathy/dysplasia (ARVC/D). Accordingly, focal RV functional abnormalities (focal akinesia, dyskinesia, or dyssynchronous RV contractility) and either RV enlargement with a gender-specific RV end-diastolic volume index (REDVI) RV systolic dysfunction are the cardiac magnetic resonance imaging (CMR)-based parameters defining the disease. See Chapter 3 for Task Force Criteria. Tissue abnormalities suggestive of inflammation, fibrosis, or fibrofatty degeneration on the standard CMR examinations were not included in the TFC due to potential variability in the interpretation of these abnormalities.

These intrinsic limitations explain the importance of the standardized ARVC/D acquisition and interpretation protocols for the diagnosis of patients with known or suspected ARVC/D. Standardized RV dimension and volume measurements [2–4] are of extreme importance. This is especially applicable to endurance athletes, who may present with borderline ECG changes [5–12]. Correct diagnosis of athletes with ARVC/D could potentially be lifesaving since the vast majority of sudden death cases in athletes with ARVC/D occur either during or soon after strenuous exercise, whereas a false-positive diagnosis could destroy their athletic careers.

Fortunately, careful evaluation of the clinical information and utilization of the ECG hemodynamic, and imaging findings can distinguish between the ARVC/D and athlete's hearts. For example, Zaidi et al. [5] compared clinical and imaging data of athletes with and without T-wave inversion and established the diagnosis of patients with ARVC/D. The authors reported the following indicators of RV pathology: history of syncope; Q waves or maximum precordial QRS amplitudes <1.8 mV; three abnormal SAECG parameters; CMR findings including delayed gadolinium enhancement, RVEF 45% or less, or RV wall motion abnormalities; >1000 ventricular extrasystoles (or >500 non-RV outflow tract ventricular beats)/24 h; and symtomatic, ventricular tachyarrhythmias, or attenuated blood pressure response during exercise.

In borderline cases, sequential CMR imaging may provide important information including: progressive worsening of RV focal and global systolic function and RV enlargement. These latter findings suggests significant RV pathology, whereas decreased RV dimensions and/or improved function following a decrease in the amount of endurance training may be consistent with the "physiologic adaptation phenomenon" observed in the athlete's heart [13].

There is evidence to suggest that normalized RV functional variables may be more useful in athletes as compared to simple indexed volumes or dimensions to distinguish the athletes heart from that of ARVC/D. Luijkx et al. [14] demonstrated that compared to RVEDVI, RVEF is more likely to distinguish ARVC/D from physiologic cardiac adaptation in athletes by CMR. They also noted that a good diagnostic alternative in athletes is the left ventricular (LV)/RV EDV ratio, representing normal proportionate adaptation of both ventricles in athletes.

CMR is a useful diagnostic modality to distinguish the athlete's heart from ARVC/D and other RV cardiomyopathies as well as RV-specific disorders (Fig. 8.1). The combination of a negative family history, absence of the significant conduction abnormalities and arrhythmias by ECG and Holter, and absence of tissue abnormalities (such as delayed hyperenhancement (DHE) or fibrofatty changes) on CMR suggests a low likelihood of ARVC/D.

Compared to echocardiography, RV focal wall motion abnormalities diagnosed by CMR is more precise due to a higher spatial resolution of CMR. However, there are several pitfalls. First, the possibility of an artifact of the observed RV wall motion abnormality on cine images should be considered (Fig. 8.2). The sequence should be repeated with change in the image settings if there is any concern of the validity of the recordings.

Suspicion of regional wall motion changes may represent normal variants and are especially prominent at the RV apex, even on 3D magnets [16]. Quick et al. noted that regional RV wall motion abnormalities are common and discussed tethering of the free RV wall by the moderator band or a complex geometry of the RV as the main reasons for this phenomenon (Fig. 8.3) [16].

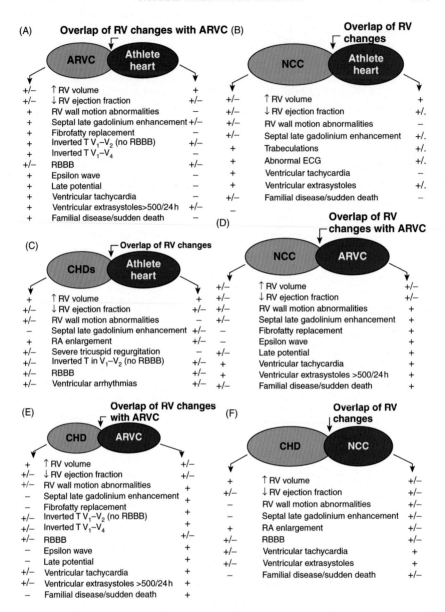

FIGURE 8.1 **Differential diagnosis of athlete's heart and some RV cardiomyopathies.** ARVC, arrhythmogenic right ventricular cardiomyopathy/dysplasia; CHD, congenital heart disease; NCC, noncompaction cardiomyopathy; RBBB, right bundle branch block. *Reproduced with permission from Ref. [15].*

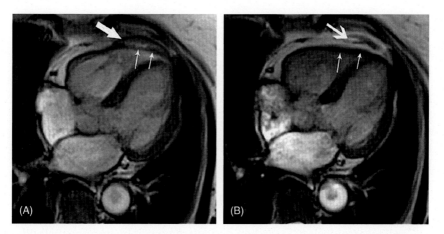

FIGURE 8.2   **Artifact mimicking a focal RV wall motion abnormality on cine CMR images.** (A) Cine SSFP four-chamber view showing an artifact at the apex of the RV (broad arrow), in front of the RV wall (small white arrows); (B) absence of the artifact in the same four-chamber view, but with a slightly different radiofrequency (open arrow). *Reproduced with permission from Ref. [3].*

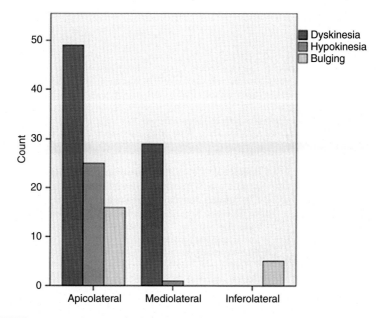

FIGURE 8.3   **Distribution of the regional RV wall motion abnormalities in the normal volunteers on 3-T CMR.** *Reproduced with permission from Ref. [16].*

They further state that "in the light of the above, one should be aware that these nonpathologic wall motion disorders can easily be mistaken for pathologic regional wall motion contraction, particularly in ARVC/D where to date, clear wall motion criteria are lacking."

It is important to note that, in the majority of ARVC/D cases, regional RV wall motion abnormalities predominantly involve basal (especially, subtricuspid) segments, RV outflow tract, and mid-free (lateral) and inferior RV walls. Involvement of the RV apex is rare in ARVC/D and is seen mostly in the advanced stages of the disease when the imaging features of the ARVC/D resemble those in dilated cardiomyopathy.

CARDIAC SARCOIDOSIS – AN ARVC/D MIMIC

Cardiac sarcoidosis (CS) can be a difficult diagnosis by imaging methods. It may require a multidisciplinary approach similar to that of the ARVC/D. Diagnostic criteria for CS are based on minor and major criteria (Table 8.1) [17] and include a combination of imaging- and nonimaging findings.

TABLE 8.1   Diagnostic Criteria for Cardiac Sarcoidosis

Histologic diagnosis group

Cardiac sarcoidosis is confirmed when endomyocardial biopsy specimens demonstrate noncaseating epithelioid cell granulomas with histologic or clinical diagnosis of extracardiac sarcoidosis.

Clinical diagnosis group

Although endomyocardial biopsy specimens do not demonstrate noncaseating epithelioid cell granulomas, extracardiac sarcoidosis is diagnosed histologically or clinically and satisfies the following conditions and more than one in six basic diagnostic criteria.
1. Two or more of the four major criteria are satisfied.
2. One of four of the major criteria and two or more of the five minor criteria are satisfied.

Major criteria	Minor criteria
1. Advanced atrioventricular block.	1. Abnormal ECG findings: ventricular arrhythmias (ventricular tachycardia, multifocal or frequent PVCs). CRBBB, axis deviation or abnormal Q wave.
2. Basal thinning of the interventricular septum.	2. Abnormal echocardiography: regional abnormal wall motion or morphologic abnormality (ventricular aneurysm, wall thickening).
3. Positive ^{67}gallium uptake in the heart.	3. Nuclear medicine: perfusion defect detected by ^{201}thallium or ^{99m}technetium myocardial scintigraphy.
4. Depressed ejection fraction of the LV (<50 %).	4. Gadolinium-enhanced CMR imaging: delayed enhancement of myocardium.
	5. Endomyocardial biopsy: interstitial fibrosis or monocyte infiltration over moderate grade.

Reproduced with permission from Ref. [18].

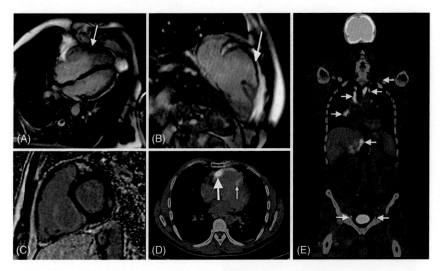

FIGURE 8.4   **Role of cardiac PET in diagnosing cardiac sarcoidosis "mimicking" the ARVC/D.** Legend from the original publication: (A) Cardiac MRI showing dilated RV (arrow) with indexed RV end-diastolic volume of 113 mL/m² and indexed RV end-systolic volume was 71 mL/m². (B) CMR demonstrates aneurysm in the inflow region (arrow) and apical hypokinesia with RVEF of 37%. (C) Cardiac MRI showing midwall-delayed gadolinium enhancement in the basal inferolateral segment of the LV. (D) Axial-fused PET/CT image demonstrates intense focal FDG uptake in the right ventricular free wall (thick arrow) and lesser uptake in the right ventricular septum (thin arrow). (E) Coronal-fused PET/CT demonstrates intense FDG uptake in enlarged neck, mediastinal, retroperitoneal, and groin nodes (arrows). Sites of bone uptake are not shown. *Reproduced with permission from Ref. [19].*

Some structural abnormalities in CS on CMR may resemble those mentioned in the TFC for the diagnostic criteria ARVC/D. Among the findings mimicking ARVC/D, there may be RV enlargement and RV dysfunction as well as regional wall motion abnormalities. Figure 8.4 demonstrates an example of a patient with sarcoid who had a positive biopsy, and RV enlargement and dysfunction (both global and regional), fulfilling a major TFC for ARVC/D [19]. In these cases, a high level of clinical suspicion and understanding of the disease pathophysiology may encourage further evaluation by cardiac positron emission tomography (PET). PET that appears to provide a high diagnostic yield in CS, to confirm myocardial inflammation and to match lesions of T2-weighted CMR imaging with areas of increased fludeoxyglucose (18F) (F18-FDG) uptake in patients with CS (Fig. 8.5) [20]. PET-CT is a powerful diagnostic tool to reveal extracardiac sarcoidosis, especially in the mediastinum/pulmonary hila. In addition to the pulmonary involvement in CS, significant LV involvement is frequently present in the CS patients. Although DHE is not specific or pathognomonic in either disorder, evidence suggests that septal scarring is more frequent in CS and is rarely found in ARVC/D (Fig. 8.6) [21].

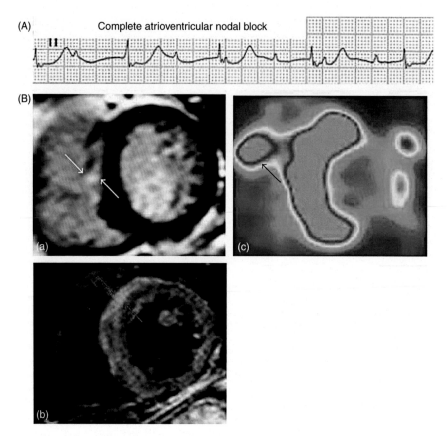

FIGURE 8.5 **Representative images of CMR and cardiac PET in a patient with pathologically proven cardiac sarcoid.** Legend from the original publication: (A, B) CMR shows areas of delayed enhancement (DE) (a) and increased T2-weighted signal (b) in the anteroseptal wall of the LV (white and red arrows). (C) 18F-FDG PET shows focal 18F-FDG uptake in the anteroseptal of the LV and RV (black arrow). *Reproduced with permission from Ref. [20].*

Other useful clinical features suggestive of CS are older age of the patient at onset, mediastinal lymphadenopathy, presentation with conduction abnormalities rather than with ventricular arrhythmias, predominant LV dysfunction, and septal wall motion abnormalities

OTHER ARVC/D MIMICS: DIAGNOSTIC DILEMMAS

Based on an analysis of a series of 657 CMR studies in patients with known or suspected ARVC/D, Quarta et al. [22] suggested the following classification of the ARVC/D mimics:

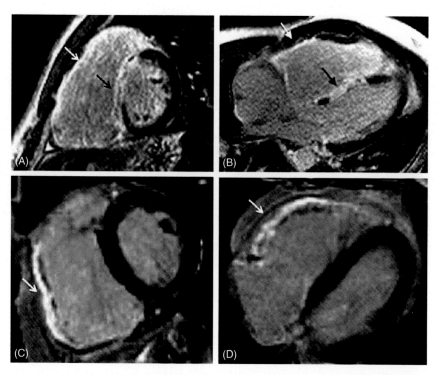

FIGURE 8.6   **Delayed enhancement CMR images in the patient with pathology-positive cardiac sarcoidosis (A, B) and ARVC/D (C, D).** White arrows depict areas of scar/fibrosis in the RV-free wall, whereas black arrows reveal areas of scar in the interventricular septum in a patient with sarcoid (not seen in patients with ARVC/D). *Reproduced with permission from Ref. [21].*

1. Cardiac displacement (pectus excavatum, scoliosis, other severe musculoskeletal problems involving the thoracic spine or rib cage; congenital absence of the pericardium)
2. RV overload (pulmonary hypertension; severe tricuspid regurgitation, atrial septal defect (ASD), etc.)
3. RV scarring (RV infarction; inflammatory cardiomyopathies, sarcoidosis; myocarditis)

Based on our clinical experience, we propose adding to this classification the RV functional abnormalities related to the pericardium and postthoracotomy changes as another ARVC/D mimic.

Each one of these mimics may have distinctive features. Clinical presentation, history, or careful evaluation of the entire imaging data may help to define the disorder (Figs 8.7–8.9). Musculoskeletal abnormalities

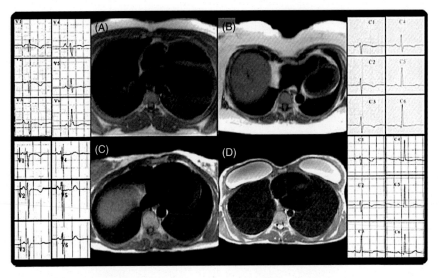

FIGURE 8.7 **Cardiac displacement mimic of the ARVC.** (A) Congenital absence of the pericardium; (B) pectus excavatum; (C) rib cage abnormality; (D) breast implants and pectus excavatum. *Reproduced with permission from Ref. [22].*

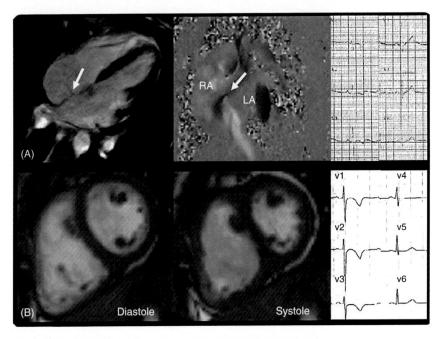

FIGURE 8.8 **RV overload mimic of the ARVC/D.** (A) Volume overload due to the secundum ASD; (B) pressure overload due to the pulmonary hypertension. *Reproduced with permission from Ref. [22].*

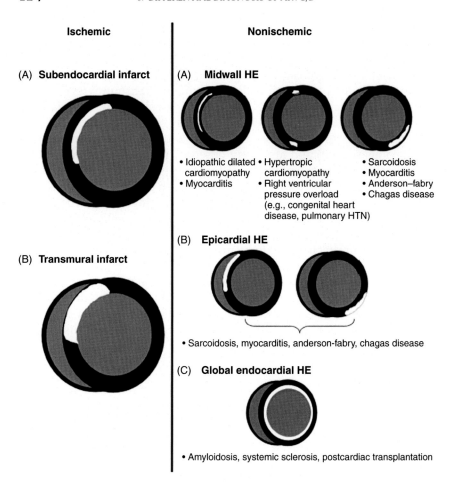

Ischemic

Nonischemic

(A) **Subendocardial infarct**

(A) **Midwall HE**

- Idiopathic dilated
  cardiomyopathy
- Myocarditis

- Hypertropic
  cardiomyopathy
- Right ventricular
  pressure overload
  (e.g., congenital heart
  disease, pulmonary HTN)

- Sarcoidosis
- Myocarditis
- Anderson–fabry
- Chagas disease

(B) **Epicardial HE**

(B) **Transmural infarct**

- Sarcoidosis, myocarditis, anderson-fabry, chagas disease

(C) **Global endocardial HE**

- Amyloidosis, systemic sclerosis, postcardiac transplantation

FIGURE 8.9   **Patterns of DHE.** *Reproduced with permission from Ref. [23].*

are relatively easily seen on CMR images. Additional comparison of the X-ray and CT scan may persist in making the correct diagnosis.

RV overload mimics include congenital heart disorders (both simple and complex congenital heart disease). It is important to review all the data available on the images. CMR is a versatile methodology and evidence of the anomalies in the great vessels or cardiac structures (patent ductus arteriosus, partial anomalous venous return, Ebstein's anomaly, unroofed coronary sinus, and sinus venosus ASD) are among the disorders that CMR may assist in diagnosing in patients with enlarged and dysfunctional RV. It is helpful to perform the phase-contrast analysis with calculation of the Qp/QS and estimates of cardiac shunting in these cases.

For patients with evidence of delayed hyperenhancement (DHE), it is important to analyze for the presence of biventricular involvement, as well as for scar in the interventricular septum, pattern of DHE in the LV: midwall (in dilated cardiomyopathies) vs. epicardial (in some inflammatory cardiomyopathies) or subendocardial to transmural (specific for myocardial infarction) (Fig. 8.9) [23]. Basal involvement, especially in the subtricuspid area, is frequent in patients with ARVC/D. Inflammatory changes are frequently seen in patients with myocarditis and CS (see the previous section) and are rarely observed in ARVC/D. Cardiac PET may increase diagnostic accuracy of the images confirming the location and pattern of inflammation seen on CMR, by an increased F18-FDG. Advanced CMR applications such as T1 and T2 mapping of the myocardium [24] can prove to be a useful addition to the standard ARVC/D protocols when myocarditis is suspected.

Finally, in our practice, we frequently observe RV focal abnormalities in patients who have had a sternotomy. This is probably due to pericardial adhesion to the posterior sternal surface. Paradoxical postoperative septal motion may also be seen in these patients. This adds to the difficulty in calculating RV volumes and RVEF, since the interventricular septum has dyssynchronous motion and may interact with the Simpson's disk method assumptions.

References

[1] Marcus FI, McKenna WJ, Sherrill D, Basso C, Bauce B, Bluemke DA, Calkins H, Corrado D, Cox MG, Daubert JP, Fontaine G, Gear K, Hauer R, Nava A, Picard MH, Protonotarios N, Saffitz JE, Sanborn DM, Steinberg JS, Tandri H, Thiene G, Towbin JA, Tsatsopoulou A, Wichter T, Zareba W. Diagnosis of arrhythmogenic right ventricular cardiomyopathy/dysplasia: proposed modification of the task force criteria. Circulation 2010;121:1533–41.

[2] Schulz-Menger J, Bluemke DA, Bremerich J, Flamm SD, Fogel MA, Friedrich MG, Kim RJ, von Knobelsdorff-Brenkenhoff F, Kramer CM, Pennell DJ, Plein S, Nagel E. Standardized image interpretation and post processing in cardiovascular magnetic resonance: Society for Cardiovascular Magnetic Resonance (SCMR) board of trustees task force on standardized post processing. J Cardiovasc Magn Reson 2013;15:35.

[3] Tavano A, Maurel B, Gaubert JY, Varoquaux A, Cassagneau P, Vidal V, Bartoli JM, Moulin G, Jacquier A. MR imaging of arrhythmogenic right ventricular dysplasia: what the radiologist needs to know. Diagn Interv Imaging 2015;96:449–60.

[4] Kramer CM, Barkhausen J, Flamm SD, Kim RJ, Nagel E. Society for Cardiovascular Magnetic Resonance Board of Trustees Task Force on Standardized Protocols. Standardized cardiovascular magnetic resonance (CMR) protocols 2013 update. J Cardiovasc Magn Reson 2013;15:91.

[5] Zaidi A, Sheikh N, Jongman JK, Gati S, Panoulas VF, Carr-White G, Papadakis M, Sharma R, Behr ER, Sharma S. Clinical differentiation between physiological remodeling and arrhythmogenic right ventricular cardiomyopathy in athletes with marked electrocardiographic repolarization anomalies. J Am Coll Cardiol 2015;65:2702–11.

[6] Major Z, Csajagi E, Kneffel Z, Kovats T, Szauder I, Sido Z, Pavlik G. Comparison of left and right ventricular adaptation in endurance-trained male athletes. Acta Physiol Hung 2015;102:23–33.

[7] Zaidi A, Ghani S, Sharma R, Oxborough D, Panoulas VF, Sheikh N, Gati S, Papadakis M, Sharma S. Physiological right ventricular adaptation in elite athletes of African and Afro-Caribbean origin. Circulation 2013;127:1783–92.

[8] Mangold S, Kramer U, Franzen E, Erz G, Bretschneider C, Seeger A, Claussen CD, Niess AM, Burgstahler C. Detection of cardiovascular disease in elite athletes using cardiac magnetic resonance imaging. Rofo 2013;185:1167–74.

[9] D'Andrea A, Riegler L, Golia E, Cocchia R, Scarafile R, Salerno G, Pezzullo E, Nunziata L, Citro R, Cuomo S, Caso P, Di Salvo G, Cittadini A, Russo MG, Calabro R, Bossone E. Range of right heart measurements in top-level athletes: the training impact. Int J Cardiol 2013;164:48–57.

[10] Wilhelm M, Roten L, Tanner H, Schmid JP, Wilhelm I, Saner H. Long-term cardiac remodeling and arrhythmias in nonelite marathon runners. Am J Cardiol 2012;110:129–35.

[11] Heidbuchel H, La Gerche A. The right heart in athletes. Evidence for exercise-induced arrhythmogenic right ventricular cardiomyopathy. Herzschrittmacherther Elektrophysiol 2012;23:82–6.

[12] Scharf M, Brem MH, Wilhelm M, Schoepf UJ, Uder M, Lell MM. Cardiac magnetic resonance assessment of left and right ventricular morphologic and functional adaptations in professional soccer players. Am Heart J 2010;159:911–8.

[13] Knackstedt C, Schmidt K, Syrocki L, Lang A, Bjarnason-Wehrens B, Hildebrandt U, Predel HG. Long-term follow-up of former world-class swimmers: evaluation of cardiovascular function. Heart Vessels 2015;30:369–78.

[14] Luijkx T, Velthuis BK, Prakken NH, Cox MG, Bots ML, Mali WP, Hauer RN, Cramer MJ. Impact of revised task force criteria: distinguishing the athlete's heart from ARVC/D using cardiac magnetic resonance imaging. Eur J Prev Cardiol 2012;19:885–91.

[15] Limongelli G, Rea A, Masarone D, Francalanci MP, Anastasakis A, Calabro R, Maria Giovanna R, Bossone E, Elliott PM, Pacileo G. Right ventricular cardiomyopathies: a multidisciplinary approach to diagnosis. Echocardiography 2015;32 Suppl. 1:75–94.

[16] Quick S, Speiser U, Kury K, Schoen S, Ibrahim K, Strasser R. Evaluation and classification of right ventricular wall motion abnormalities in healthy subjects by 3-tesla cardiovascular magnetic resonance imaging. Neth Heart J 2015;23:64–9.

[17] Diagnostic guideline and criteria for sarcoidosis – 2006. Nippon Ganka Gakkai Zasshi 2007;111:117–121.

[18] Skali H, Schulman AR, Dorbala S. 18F-FDG PET/CT for the assessment of myocardial sarcoidosis. Curr Cardiol Rep 2013;15:352.

[19] Choo WK, Denison AR, Miller DR, Dempsey OJ, Dawson DK, Broadhurst PA. Cardiac sarcoid or arrhythmogenic right ventricular cardiomyopathy: a role for positron emission tomography (PET)? J Nucl Cardiol 2013;20:479–80.

[20] Orii M, Imanishi T, Akasaka T. Assessment of cardiac sarcoidosis with advanced imaging modalities. Biomed Res Int 2014;2014:897956.

[21] Philips B, Madhavan S, James CA, te Riele AS, Murray B, Tichnell C, Bhonsale A, Nazarian S, Judge DP, Calkins H, Tandri H, Cheng A. Arrhythmogenic right ventricular dysplasia/cardiomyopathy and cardiac sarcoidosis: distinguishing features when the diagnosis is unclear. Circ Arrhythm Electrophysiol 2014;7:230–6.

[22] Quarta G, Husain SI, Flett AS, Sado DM, Chao CY, Tome Esteban MT, McKenna WJ, Pantazis A, Moon JC. Arrhythmogenic right ventricular cardiomyopathy mimics: role of cardiovascular magnetic resonance. J Cardiovasc Magn Reson 2013;15:16.

[23] Mahrholdt H, Wagner A, Judd RM, Sechtem U, Kim RJ. Delayed enhancement cardiovascular magnetic resonance assessment of non-ischaemic cardiomyopathies. Eur Heart J 2005;26:1461–74.

[24] Hagio T, Huang C, Abidov A, Singh J, Ainapurapu B, Squire S, Bruck D, Altbach MI. T2 mapping of the heart with a double-inversion radial fast spin-echo method with indirect echo compensation. J Cardiovasc Magn Reson 2015;17:24.

9

Special Cases and Special Populations: Tips and Tricks to Obtain a Diagnostic CMR

Isabel B. Oliva, Aiden Abidov

Department of Medicine/Division of Cardiology
and Department of Medical Imaging,
University of Arizona, Tucson, AZ, USA

CMR IN PATIENTS WITH PERMANENT PACEMAKERS AND ICDs

Many patients with ARVC/D develop significant ventricular arrhythmias, or unfortunately present with sudden death. Based on current evidence, ICD placement is a part of the risk management in ARVC/D, making follow-up CMR a challenge.

There is a growing list of scientific papers supporting the feasibility and safety of CMR imaging in patients with pacemakers/ICDs. Here are just a few examples. Luechinger et al. [1] found no evidence of irreversible damage to the pacemaker or its components when imaging these patients in 1.5-T scanners. Roguin et al. [2] and Naehle et al. [3] have documented that MRI is safe with regard to radiofrequency-related heating of the lead tip.

With clinical protocols developed as a successful product of collaboration of the imaging labs and the pacemaker clinics, safety of the patient with an ICD who is not pacemaker-dependent, is not a problem. What may be of concern is the image quality and especially, potentially significant image degradation of the right ventricle (due to lead artifact) and anteroseptal wall image degradation secondary to the generator artifact (Fig. 9.1).

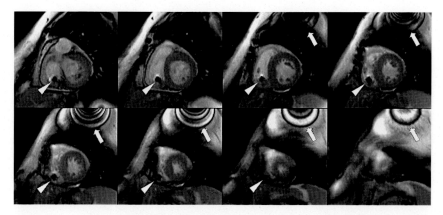

FIGURE 9.1 **Typical pacemaker-related imaging artifacts on CMR.** Metal artifact from the pacemaker generator (arrows) and the pacemaker lead (arrowheads) is present. *Reproduced with permission from Ref. [4].*

At our institution we have a protocol for scanning patients with pacemaker/ICDs. The most important component of this protocol is our multidisciplinary approach. As a part of our protocol, the electrophysiology (EP) team sees all patients prior to any MR image acquisition.

Patient selection criteria for the pacemaker-CMR protocol includes cardiac lead placement more than 6 months prior to the imaging study; stable pace, sensing, and impedance parameters, normal battery function, and stable unpaced heart rhythm with HR > 50 bpm (nonpacer dependent). The patient is determined unsafe for MRI if he/she does not meet all of these criteria.

When MR imaging is determined to be safe for the patient, an EP lab technician comes to the MR scanner and interrogates the pacemaker prior to the patient entering the scanner room. The device is programmed to sensing only, no pacing. For patients with an ICD, detection of ventricular tachycardia and fibrillation is turned off.

A diagnostic ECG is performed before and after the test. Patients are scanned with continuous monitoring by a registered nurse of their ECG and pulse oximetry. A physician is always at the scanner and resuscitation equipment is immediately available by the MR suite. Collaboration with the medical physicist is another essential factor permitting optimization of protocols.

Despite the inclusion of ferromagnetic material within the field of view (FOV), diagnostic quality images can be obtained when using the appropriate sequences. Whole-body specific absorption rate (SAR) is limited to 1.5 W/kg. When acquiring cine images using the steady-state free precession (SSFP) technique, banding artifacts (signal voids) are seen extending beyond the device generator because it is a ferromagnetic object. This artifact can be minimized with the use of spoiled gradient recalled echo

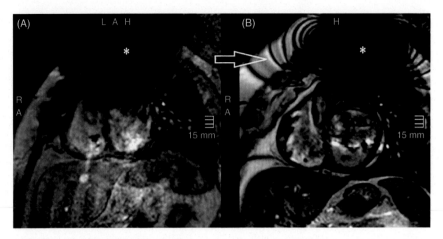

FIGURE 9.2   **Metal artifacts in patients with ICD.** (A) On GRE images, asterisk shows a dark signal – typical susceptibility artifact surrounding the ICD generator. (B) On SSFP images, asterisk shows typical dark banding artifacts surrounding the ICD generator and extending well beyond the device. *Reproduced with permission from Ref. [3].*

(GRE) sequence (Fig. 9.2) [3]. The limitation is that the images have lower contrast-to-noise ratio compared to SSFP. Shimming is also used to reduce metal-related artifacts since it corrects the inhomogeneity of the magnetic field related to the ferromagnetic components of the cardiac device. We routinely use autoshimming but an additional dedicated cardiac shimming can be done for these cases. Other strategies that can be used to reduce metal-related artifacts include increasing the bandwidth and resolution, reducing slice thickness, and swapping the read and phase encoding directions.

After the image acquisition is completed, the EP technician interrogates the cardiac device again and all settings are reset. Lead impedance as well as sensing and pacing characteristics are confirmed to be stable. Typically, a follow-up visit is scheduled for these patients in the pacemaker clinic in 3 months following the CMR study.

ARRHYTHMIAS

The images acquired during a cardiac MRI examination are routinely synchronized with cardiac motion; therefore, accurate synchronization of data acquisition is essential to obtain motion-free (nonblurred) images in a timely manner.

The ECG provides physiologic data for the cardiac gating algorithms. When using ECG gating, the timing of image acquisition is controlled by the cardiac cycle. Normally, to obtain cine loop of one image of the heart, the data are collected with multiple heartbeats to fill separate areas of the k-space. In the presence of arrhythmias, data acquisition is usually inaccurately synchronized to the ECG so the resulting images demonstrate motion artifacts and may be nondiagnostic. This is particularly important for patients with suspected ARVC/D, as a blurred image decreases the sensitivity to detect subtle wall motion abnormalities and may result in under- or overestimation of the right ventricular end-diastolic volume.

A common determinant of poor ECG gating is suboptimal electrode placement. The electrodes should be positioned 4–6 in. apart and must have excellent contact with the skin. The data obtained during CMR examination are triggered by the R wave, indicating the importance of good amplitude or steep slope of the R wave.

Peripheral pulse monitoring can be used to gate the data acquisition but it commonly results in lack of data acquisition during systole in cases of prospective gating. When available, retrospective gating is preferable while using peripheral pulse gating.

Patients with ARVC/D may have frequent ventricular arrhythmias. Premature beats are a problem as the data to be acquired during diastole will be acquired during systole. Most MR scanners have the arrhythmia detection software that allows narrowing of the acquisition window in order to avoid unwanted beats.

In cases of frequent arrhythmias, real-time imaging can be performed. This technique permits rapid acquisition of the entire data during one heartbeat and it can obviate ECG gating. The shortcoming of this approach is the somewhat low-temporal resolution and decreased signal-to-noise ratio (SNR) of the resulting images.

INABILITY TO BREATH HOLD

It is not uncommon to encounter patients who are not able to hold their breath for 10–15 s. The reasons are variable and include cardiopulmonary diseases, anxiety, and sedation. The image acquisition time is determined mainly by the patient's heart rate and regularity. Slower heart rate and irregular rhythm result in longer acquisition times.

The number of concatenations also influences acquisition time and duration of breath holds. Increasing concatenations will decrease the duration of breath hold but it will also increase acquisition time. Additionally, some sequences do not allow any change in the number of concatenations. Other strategies to decrease the time of breath hold include decreasing phase oversampling, but that results in lower resolution.

For extreme cases, images can be acquired without breath hold (free breathing) but one must increase the number of averages to three or four so that the respiratory motion artifact is minimized.

We frequently scan patients who are not capable of breath holding; accordingly, our imaging protocols include nonbreath-hold sequences for all the key elements of the CMR study: from the real-time nonbreath-hold cine images for chamber volumetric measurements to the nonbreath-hold delayed enhancement sequences and phase-contrast imaging. Even nonbreath-hold CMR images are typically better than standard echocardiographic images and can be effectively utilized for functional assessment. Delayed enhancement images are invaluable in assessing the presence and amount of the myocardial scar/fibrosis and are unique for the CMR modality. Similarly, phase-contrast imaging provides unique hemodynamic data. These methods should always be considered and discussed with the patient when the information obtained and expected clinical advantages of the CMR are considerably higher compared to the level of the patient's discomfort and any possible associated risk.

FIELD OF VIEW

Signal to noise ratio (SNR) and spatial resolution are closely related to the FOV because the latter determines the size of the pixels when the matrix size is held constant. Pixel size determines the spatial resolution of the final image.

Obese patients have a larger thoracic cage, so as to avoid aliasing (wraparound artifact), a larger FOV is required when imaging their thorax. There is an "associated cost" for this maneuver. Increasing the FOV will lead to larger but fewer pixels and therefore lower spatial resolution although higher SNR. To maintain the pixel size one might consider increasing the matrix size but that will lead to an increase in image acquisition time.

PEDIATRIC PATIENTS

Imaging pediatric patients is a special problem because they usually are small and unable to breath hold. The patient's smaller size requires a smaller FOV but that may result in lower SNR. The best strategy is to reduce the FOV in the phase-encoding direction only (rectangular FOV); this approach will maintain the original spatial resolution since this is determined by the matrix size in the frequency encoding direction. One must be careful not to reduce the FOV too much in the phase-encoding direction as this may result in a wraparound artifact. The phase-encoding direction should always be aligned with the patient's smallest dimension (usually anteroposterior) and there are some antialiasing options that can be used such as "no phase wrap."

To obtain appropriate cardiac MR images in pediatric patients, the previously discussed approach of scanning without breath hold is preferable. Generally, in very young pediatric patients we recommend feeding immediately prior to image acquisition followed by swaddling and scanning without breath hold using three to four averages.

Older pediatric patients may be able to breath hold appropriately but the noise when inside the MR scanner may scare them. Thus, we provide goggles and headphones so they can watch their favorite movie while we acquire the images. One of the parents is usually present in the scanner room with the patient, so the patients feel more relaxed.

References

[1] Luechinger R, Duru F, Scheidegger MB, Boesiger P, Candinas R. Force and torque effects of a 1.5-Tesla MRI scanner on cardiac pacemakers and ICDs. Pacing Clin Electrophysiol 2001;24:199–205.
[2] Roguin A, Zviman MM, Meininger GR, et al. Modern pacemaker and implantable cardioverter/defibrillator systems can be magnetic resonance imaging safe: *in vitro* and *in vivo* assessment of safety and function at 1.5 T. Circulation 2004;110:475–82.
[3] Naehle CP, Kreuz J, Strach K, et al. Safety, feasibility, and diagnostic value of cardiac magnetic resonance imaging in patients with cardiac pacemakers and implantable cardioverters/defibrillators at 1.5 T. Am Heart J 2011;161:1096–105.
[4] Ferreira AM, Mendes L, Soares L, da Graca Correia M, Gil V. Cardiac magnetic resonance in a patient with MRI-conditional pacemaker. Rev Port Cardiol 2013;32:159–62.

Prognostic Value of Cardiac MRI in ARVC/D

Isabel B. Oliva, Aiden Abidov

Department of Medicine/Division of Cardiology
and Department of Medical Imaging,
University of Arizona, Tucson, AZ, USA

Arrhythmogenic right ventricular cardiomyopathy/dysplasia (ARVC/D) is an inherited autosomal dominant cardiomyopathy characterized by the predominance of ventricular arrhythmias [1]. In the previous chapters, we described diagnostic challenges in patients with suspected ARVC/D. This is a clinically heterogeneous disease due to its incomplete penetrance and variable expression. Symptoms related to ARVC/D range from palpitation, fatigue, and chest pain, to more severe presentations such as lightheadedness and near syncope due to rapid monomorphic ventricular tachycardia (VT), as well as heart failure, and sudden cardiac death (SCD). However, many patients are asymptomatic, especially in the early stages of disease.

Fibrofatty replacement of the RV myocardium interferes with electrical impulse conduction. The electrical instability of this dystrophic myocardium can precipitate sudden death. Progressive myocardial cell loss also results in ventricular dysfunction and heart failure. Both these factors influence the natural history of ARVC/D [2–5]. Areas of myocardial fibrosis may lead to reentry, which can cause hemodynamically stable or unstable ventricular arrhythmias in patients with relatively stable disease [6–8]. Unfortunately, in ARVC/D patients, the very first clinical presentation is SDC.

What is the magnitude of the problem? ARVC/D has prevalence estimated between 1:2000 and 1:5000 of the general population and is a cause of SCD in youth [9]. ARVC accounts for 3–4% of deaths in sports and 5% of SCD in people younger than 65 years of age [9,10]. SCD can occur at any time during the course of the disease with an incidence varying from 0.1% to 3% per year [11]. Evidence suggests that SCD is caused by hemodynamically unstable VT or ventricular fibrillation (VF). SCD in young

Cardiac MRI in the Diagnosis, Clinical Management and Prognosis of Arrhythmogenic Right Ventricular Cardiomyopathy/Dysplasia

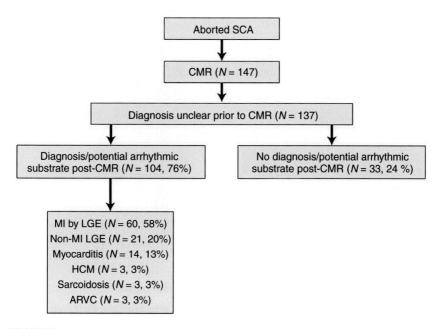

FIGURE 10.1   **CMR-based diagnosis of the potential arrhythmic source among survivors of SCD.** *Reproduced with permission from Ref. [13].*

asymptomatic patients is usually due to VF that occurs during the active phase of acute myocyte death and reactive inflammation [6–8]. Thus, appropriate screening and risk stratification is pivotal in patients with ARVC/D due to the associated incidence of SCD.

CMR findings and evidence of electrical instability allow initial screening and selection of patients at expected high risk among those patients with known or suspected ARVC/D (Figs 10.1 and 10.2) [12, 13].

In this regard, presence of the unique CMR feature – delayed gadolinium hyperenhancement (DHE) is associated with incremental diagnostic and prognostic value in ARVC/D. It allows identification of the patients at higher risk among survivors of aborted cardiac arrest and reveals an arrhythmic substrate in 76% of the cases (Fig. 10.1) [13]. Risk stratification of patients at a high risk of major adverse cardiac events (MACE) based on the amount of DHE is effective not only in ARVC/D but also in other cardiac disorders leading to arrhythmias; the threshold of DHE >8% is clearly associated with increased MACE risk (Fig. 10.3) [13].

The natural history of ARVC/D includes four phases. The initial phase is subclinical with subtle structural changes commonly located in the so-called triangle of dysplasia. Despite being asymptomatic these patients are at increased risk of SCD, particularly during exercise [14]. In the second

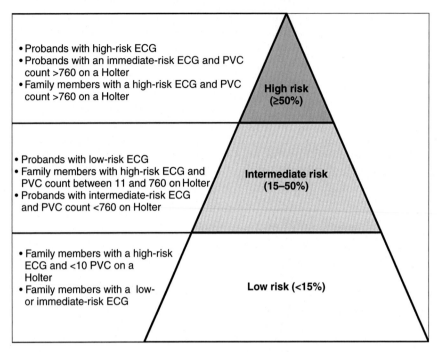

- Probands with high-risk ECG
- Probands with an immediate-risk ECG and PVC count >760 on a Holter
- Family members with a high-risk ECG and PVC count >760 on a Holter

High risk (≥50%)

- Probands with low-risk ECG
- Family members with high-risk ECG and PVC count between 11 and 760 on Holter
- Probands with intermediate-risk ECG and PVC count <760 on Holter

Intermediate risk (15–50%)

- Family members with a high-risk ECG and <10 PVC on a Holter
- Family members with a low- or immediate-risk ECG

Low risk (<15%)

FIGURE 10.2 Arrhythmic risk stratification among patients with ARVC/D-associated desmosomal mutations based on presentation, ECG, and Holter monitoring. *Reproduced with permission from Ref. [12].*

phase, patients start developing symptomatic ventricular arrhythmias and demonstrate more apparent morphological and functional changes in the right ventricle. The third phase is characterized by diffuse myocardial damage with preserved LV function. Diffuse biventricular involvement occurs in the advanced phase with a phenotype resembling that of dilated cardiomyopathy [5,15,16].

Mutation-specific genetic testing is recommended for family members following the identification of an ARVC/D-causative mutation in an index case [17]. Many of these patients detected by genetic testing are in the initial subclinical phase of the disease. Despite recent developments, the diagnosis of ARVC remains difficult, particularly in asymptomatic patients who may be at increased risk of SCD [6]. Cardioverter defibrillator (ICD) implantation is mandatory in patients with a history of cardiac arrest or hemodynamically unstable VT because its value for secondary prevention of SCD is well established [18]. ICD therapy is also indicated in patients with history of premature SCD in one or more first-degree relatives. Multiple studies have shown that ICD implantation changes the natural history of ARVC/D and improves survival in patients

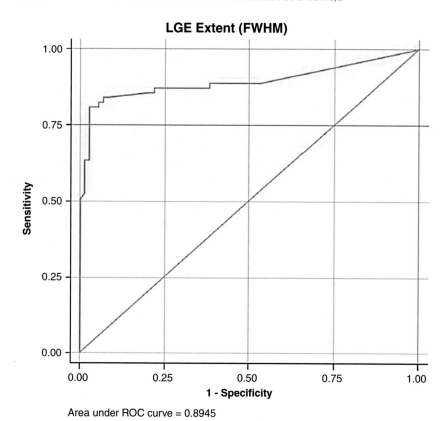

Area under ROC curve = 0.8945

FIGURE 10.3   **ROC curve for prediction of MACE among survivors of SCD.** The cutoff of DHE with the highest sensitivity and specificity (8.1%) was utilized. *Reproduced with permission from Ref. [13].*

with a high-risk profile. Despite recent progress in understanding the pathogenesis and genetics of this disease, the role of ICD implantation in preventing SCD in asymptomatic mutation carriers remains controversial. Even in patients with an ICD, the presence of "electrical" risk markers (such as inducible arrhythmias on EP study, etc.) was associated with adverse outcomes (Fig. 10.4) [19].

What about prognostic evaluation in ARVC/D patients with subclinical disease? Early detection of concealed forms is essential for timely implementation of preventive strategies [14]. CMR has markedly evolved over the past 10 years and its use to diagnose and characterize right heart pathology is well established. The variable shape of the right ventricle and poor anatomical windows are limiting factors in the echocardiogram assessment of the right ventricle. CMR allows accurate as-

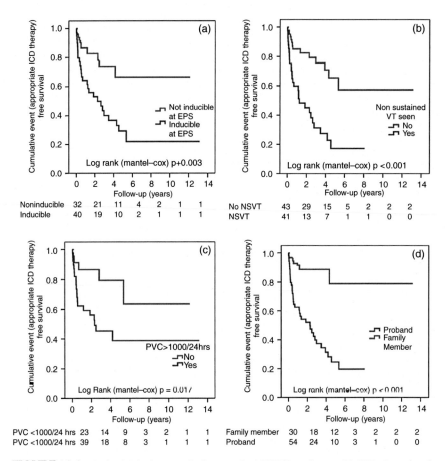

FIGURE 10.4 **Arrhythmia-free survival curves in ARVC/D patients with ICD.** *Reproduced with permission from Ref. [19].*

sessment of the right ventricular size and function in addition to precise characterization of the left ventricle and is currently the gold-standard diagnostic test when evaluating the right ventricle. CMR has the potential to detect regional and diastolic right ventricular dysfunction, which may represent an early manifestation of ARVC/D, and accordingly, yield more diagnostic information to detect early and concealed cases of ARVC, which could otherwise remain undiagnosed by the Task Force criteria (TFC) [14]. Recently published data suggest that use of advanced technology, such as strain imaging or various automated detection algorithms of wall motion abnormality, leads to improved diagnosis and stratification of patients with early disease, before the right ventricular abnormalities are noted [20].

CMR plays an important role in verifying the diagnosis as well as risk stratifying patients into groups with increased clinical risk of SCD versus groups with a low risk who do not require intervention.

However, the role of CMR is not limited by the baseline, initial diagnosis of ARVC/D. Regardless of the severity and the clinical phase of the disease (early vs. advanced), an important question is what is the best diagnostic approach to follow these patients? How soon after the initial diagnosis of ARVC/D, should we rescan the patient?

Current guidelines recommend serial screening of genetically predisposed patients using a combination of ECG, Holter monitoring, and imaging [17]. Several studies have focused on identifying parameters that can be used to determine the risk of SCD. The role of ECG is well established when stratifying patients with ARVC/D. te Riele et al. studied 69 patients with ARVC/D associated pathogenic mutations and without prior sustained ventricular arrhythmias and assessed the association between abnormal electrical test results and CMR abnormalities [21]. They found that electrical abnormalities on ECG and Holter monitoring preceded detectable structural abnormalities in ARVC/D desmosomal mutation carriers, suggesting that CMR probably does not add prognostic value in the absence of baseline electrical abnormalities (see Fig. 3.5, Chapter 3). On the other hand, the presence of both electrical and CMR abnormalities identify patients at high risk who might benefit from prophylactic ICD placement, since sustained ventricular arrhythmias occur exclusively in patients with abnormal CMR results [21].

QRS prolongation and dispersion correlate with the arrhythmic risk and predict SCD. QRS dispersion represents regional inhomogeneity of depolarization, as a consequence of ventricular conduction defects. Turrini et al. found that QRS dispersion ≥ 40 ms was the strongest independent predictor of SCD in patients with ARVC/D [23]. Ma et al. analyzed the relationship of the CMR findings and QRS dispersion and found a significant correlation between the QRS dispersion≥ 40 ms and right ventricular end-diastolic volume, right ventricular end-systolic volume, and right ventricular outflow tract (RVOT) area [14]. The dimension of the RVOT is an echocardiographic criterion for ARVC/D but this finding is not included in the MRI assessment in the modified Task Force criteria [24]. The presence of myocardial fibrosis on CMR was more prevalent in patients with QRS dispersion ≥ 40 ms compared to QRS dispersion < 40 ms. There was no significant correlation between QRS dispersion and fibro-fatty infiltration of the myocardium.

CMR permits tissue characterization with delayed contrast enhancement enabling the diagnosis of scar tissue in both right and left ventricles. Even though CMR can detect myocardial scar in the right ventricle, the sensitivity of this test remains low. The thin wall of the right ventricle makes quantitative analyses of delayed enhancement difficult. Marra et al. found a mismatch in the right ventricle between EVM (RV-EVM) and CMR (RV-CMR) when assessing the presence of RV fibrosis with fewer lesions detected by CMR compared to EVM [25]. However, patients with abnormal RV-EVM and normal RV-CMR were found to have a high prevalence of scar in the left ventricle, indicating the diagnostic relevance of LV scar detection by CMR.

Initial CMR studies documented the presence of high T1 signal within the RV myocardium in up to 75% of patients with ARVC/D compared to none in the control group [26]. Keller et al. found good diagnostic agreement between CMR and traditional diagnostic tests for ARVC/D and demonstrated that a negative CMR study was associated with arrhythmia-free survival. This study found that fatty infiltration had the highest diagnostic value of all tested CMR parameters and that functional analysis of the RV free wall provided important additional information [27]. Since the right ventricular wall is thin, the detection of fatty deposition within the dystrophic myocardium is difficult and unreliable. Based on this lower reproducibility the revised TFC does not include fatty infiltration as a criterion [24]. Schick et al. have recently shown that chemical shift selective sequences improve detection of fat and fibrosis within the RV myocardium [28], which may be more reliable and reproducible for tissue characterization.

DHE appears to be more reproducible and easier to interpret compared with fatty infiltration and even though this criterion was removed from the modified TFC. We believe there is a benefit in performing late gadolinium enhancement imaging in all patients with suspected or confirmed ARVC/D. Saranathan and coworkers found that the presence of late gadolinium enhancement was associated with inducible ventricular stimulation in patients with documented ARVC/D [29]. ARVC/D is currently recognized as a biventricular disease [30]. Hulot et al. found that clinical signs of RV failure and LV dysfunction without VT were independently associated with cardiac death [11]. Recently, Pinamonti et al. confirmed that LV dysfunction influences the long-term outcome of these patients [31], a finding previously observed by others [11,32]. te Riele et al. found that approximately 64% of mutation carriers who have biventricular involvement by CMR, experienced sustained tachyarrhythmias [21].

The original TFC stated that only mild LV dysfunction should be present in patients with ARVC/D with greater LV involvement usually occurring in the later stages of the disease. Current studies have found that LV involvement is variable. Advanced biventricular and early LV involvement have been identified in some cohorts [5,33]. Paetsch et al. recently demonstrated a left-dominant type of arrhythmogenic cardiomyopathy involving the posterolateral region of the left ventricle [34]. Pinamonti et al. reported a case of a young patient with biventricular involvement with a postmortem specimen revealing fatty infiltration in the myocardium of both right and left ventricles [35].

The 2010 TFC defined a diagnostic consensus of a combination of several parameters including morphology, histology, electrical characteristics, and clinical presentation [24]. These factors were weighted into minor and major criteria according to their relative significance. Advancements in genetics and imaging have led to a better understanding of this disease.

Deac et al. analyzed the prognostic value of CMR in 369 patients with at least one criterion for ARVC/D and found that an abnormal CMR study was a significant predictor of an end-point composite that included cardiac death, sustained VT, VF, and appropriate ICD discharge. The risk of an abnormal CMR was compared to traditional clinical parameters. After multivariate analysis only abnormal CMR and VT remained significant predictors of outcomes [36]. This recent study corroborates the findings of the previous work of Aquaro et al. who found similar results. Patients with an abnormal CMR were at increased risk of cardiac events [37].

CMR is an excellent technique to evaluate patients with ARVC/D because of its ability to detect anatomical, functional, and tissue-specific abnormalities. Even though myocardial delayed enhancement and the presence of a dilated and dysmorphic RVOT are not part of the revised TFC, these factors should be assessed when imaging patients with suspected ARVC/D. We think that CMR should be included in the diagnostic evaluation when risk-stratifying patients who have a pathogenic mutation and are at risk of SCD; however, the diagnosis and evaluation of ARVC/D should not be based solely on any one test. Availability of the scanner and clinical experience of the readers are important considerations when implementing this imaging modality.

DETERMINING THE NEED AND OPTIMAL TIMING FOR FOLLOW-UP CMR

Incomplete penetrance and variable expressivity complicate screening of family members of patients with ARVC/D. There is a known risk of ventricular arrhythmias in patients diagnosed with ARVC/D but this arrhythmic propensity may not be present in the at-risk relatives. Current literature is controversial regarding the optimal approach to longitudinal noninvasive follow-up of at-risk family members of ARVC/D probands, particularly with regard to repeat MR imaging.

CMR is currently the imaging modality of choice when evaluating for structural changes. In a recent prospective study, te Riele et al. studied 117 relatives at risk of developing ARVC/D. Seventy-four subjects did not have a diagnosis of ARVC/D at initial evaluation and were followed for 4 years. At first evaluation these subjects had a low short-term risk of sustained arrhythmia. Almost one-third of at-risk relatives had electrical progression. However, structural progression by CMR was seen in only one patient and this patient had an interval increase in RV end-diastolic volume and decrease in RV ejection fraction (RVEF), fulfilling a minor 2010 TFC for CMR at last evaluation. This patient who had structural progression had an abnormal ECG with precordial T wave inversion at enrollment. This observation adds to the increasing evidence that electrical abnormalities are more prevalent and precede detectable structural changes on CMR (Fig. 10.5) [22].

Protonotarios et al. also showed that electrical abnormalities precede structural changes [38]. This group followed 205 at-risk subjects for a median of 4 years and diagnosed 16 mutation carriers with ARVC/D. All newly diagnosed patients had ECG abnormalities and only 31% had structural abnormalities.

Another important and yet unanswered question is: how often should relatives of ARVC/D probands be screened? te Riele et al. found a slow rate of progression of disease with interval change usually not appreciated on short-term follow-up studies [22]. These results concur with prior evidence that showed minimal interval change in electroanatomic scar mapping and CMR during a follow-up period of approximately 4 years

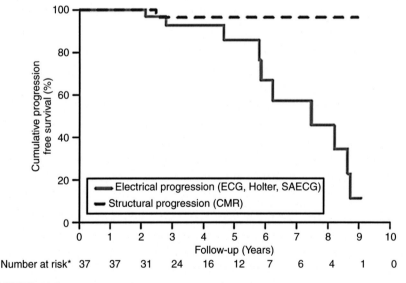

FIGURE 10.5 **Time for ARVC/D progression among patients with complete reevaluation.** *Reproduced with permission from Ref. [22].*

[39,40]. Changes in short-term follow-up are likely minor, and should be taken into consideration when reevaluating family members of the ARVC/D index patients at 2–3 years interval.

CMR is an important prognostic tool and an excellent modality for risk stratification of patients with known or suspected ARVC/D regardless of the clinical phase of the disease. Using advanced methodology, CMR may reveal subtle changes in right ventricular function, geometry, and structure well before the RVEF is decreased. In patients with phenotypical positivity, CMR may be helpful to establish the diagnosis and provide reliable longitudinal follow-up. Since electrical abnormalities precede detectable structural changes, the use of serial CMR screening may be restricted to symptomatic family members with electrical abnormalities on ECG and Holter monitoring. CMR has excellent spatial resolution and good inter- and intraobserver variability and is an excellent modality for sequential imaging. Based on published literature, the majority of the patients should be rescanned every 4 years, unless new symptoms develop.

References

[1] Hauer RNW, Marcus FI, Cox MGJP, 2013. Arrhythmogenic right ventricular dysplasia/cardiomyopathy. In: Chatterjee K, Abboud F, Anderson M, Heistad D, Kerber R, editors. Cardiology – an illustrated text book. London: JP Medical LTD. p. 705–716.

[2] Corrado D, Basso C, Thiene G. Arrhythmogenic right ventricular cardiomyopathy: an update. Heart 2009;95:766–73.

[3] Basso C, Thiene G, Corrado D, Angelini A, Nava A, Valente M. Arrhythmogenic right ventricular cardiomyopathy. Dysplasia, dystrophy, or myocarditis? Circulation 1996;94:983–91.

[4] Dalal D, Nasir K, Bomma C, Prakasa K, Tandri H, Piccini J, Roguin A, Tichnell C, James C, Russell SD, Judge DP, Abraham T, Spevak PJ, Bluemke DA, Calkins H. Arrhythmogenic right ventricular dysplasia: a United States experience. Circulation 2005;112:3823–32.

[5] Corrado D, Basso C, Thiene G, McKenna WJ, Davies MJ, Fontaliran F, Nava A, Silvestri F, Blomstrom-Lundqvist C, Wlodarska EK, Fontaine G, Camerini F. Spectrum of clinicopathologic manifestations of arrhythmogenic right ventricular cardiomyopathy/dysplasia: a multicenter study. J Am Coll Cardiol 1997;30:1512–20.

[6] Basso C, Corrado D, Marcus FI, Nava A, Thiene G. Arrhythmogenic right ventricular cardiomyopathy. Lancet 2009;373:1289–300.

[7] Corrado D, Basso C, Pilichou K, Thiene G. Molecular biology and clinical management of arrhythmogenic right ventricular cardiomyopathy/dysplasia. Heart 2011;97:530–9.

[8] Corrado D, Leoni L, Link MS, Della Bella P, Gaita F, Curnis A, Salerno JU, Igidbashian D, Raviele A, Disertori M, Zanotto G, Verlato R, Vergara G, Delise P, Turrini P, Basso C, Naccarella F, Maddalena F, Estes NA III, Buja G, Thiene G. Implantable cardioverter-defibrillator therapy for prevention of sudden death in patients with arrhythmogenic right ventricular cardiomyopathy/dysplasia. Circulation 2003;108:3084–91.

[9] Thiene G, Nava A, Corrado D, Rossi L, Pennelli N. Right ventricular cardiomyopathy and sudden death in young people. N Engl J Med 1988;318:129–33.

[10] Peters S, Peters H, Thierfelder L. Risk stratification of sudden cardiac death and malignant ventricular arrhythmias in right ventricular dysplasia-cardiomyopathy. Int J Cardiol 1999;71:243–50.

[11] Hulot JS, Jouven X, Empana JP, Frank R, Fontaine G. Natural history and risk stratification of arrhythmogenic right ventricular dysplasia/cardiomyopathy. Circulation 2004;110:1879–84.

[12] Bhonsale A, James CA, Tichnell C, Murray B, Madhavan S, Philips B, Russell SD, Abraham T, Tandri H, Judge DP, Calkins H. Risk stratification in arrhythmogenic right ventricular dysplasia/cardiomyopathy-associated desmosomal mutation carriers. Circ Arrhythm Electrophysiol 2013;6:569–78.

[13] Neilan TG, Farhad H, Mayrhofer T, Shah RV, Dodson JA, Abbasi SA, Danik SB, Verdini DJ, Tokuda M, Tedrow UB, Jerosch-Herold M, Hoffmann U, Ghoshhajra BB, Stevenson WG, Kwong RY. Late gadolinium enhancement among survivors of sudden cardiac arrest. JACC Cardiovasc Imaging 2015;8:414–23.

[14] Ma N, Cheng H, Lu M, Jiang S, Yin G, Zhao S. Cardiac magnetic resonance imaging in arrhythmogenic right ventricular cardiomyopathy: correlation to the QRS dispersion. Magn Reson Imaging 2012;30:1454–60.

[15] Sen-Chowdhry S, Lowe MD, Sporton SC, McKenna WJ. Arrhythmogenic right ventricular cardiomyopathy: clinical presentation, diagnosis, and management. Am J Med 2004;117:685–95.

[16] Corrado D, Fontaine G, Marcus FI, McKenna WJ, Nava A, Thiene G, Wichter T. Arrhythmogenic right ventricular dysplasia/cardiomyopathy: need for an international registry. Study Group on Arrhythmogenic Right Ventricular Dysplasia/Cardiomyopathy of the Working Groups on Myocardial and Pericardial Disease and Arrhythmias of the European Society of Cardiology and of the Scientific Council on Cardiomyopathies of the World Heart Federation. Circulation 2000;101:E101–6.

[17] Ackerman MJ, Priori SG, Willems S, Berul C, Brugada R, Calkins H, Camm AJ, Ellinor PT, Gollob M, Hamilton R, Hershberger RE, Judge DP, Le Marec H, McKenna WJ, Schulze-Bahr E, Semsarian C, Towbin JA, Watkins H, Wilde A, Wolpert C, Zipes DP. HRS/EHRA expert consensus statement on the state of genetic testing for the channelopathies and cardiomyopathies: this document was developed as a partnership between the Heart Rhythm Society (HRS) and the European Heart Rhythm Association (EHRA). Europace 2011;13:1077–109.

[18] Wichter T, Paul TM, Eckardt L, Gerdes P, Kirchhof P, Bocker D, Breithardt G. Arrhythmogenic right ventricular cardiomyopathy. Antiarrhythmic drugs, catheter ablation, or ICD? Herz 2005;30:91–101.

[19] Calkins H. Arrhythmogenic right ventricular dysplasia/cardiomyopathy-three decades of progress. Circ J 2015;79:901–13.

[20] Heermann P, Hedderich DM, Paul M, Schulke C, Kroeger JR, Baessler B, Wichter T, Maintz D, Waltenberger J, Heindel W, Bunck AC. Biventricular myocardial strain analysis in patients with arrhythmogenic right ventricular cardiomyopathy (ARVC) using cardiovascular magnetic resonance feature tracking. J Cardiovasc Magn Reson 2014;16:75.

[21] te Riele AS, Bhonsale A, James CA, Rastegar N, Murray B, Burt JR, Tichnell C, Madhavan S, Judge DP, Bluemke DA, Zimmerman SL, Kamel IR, Calkins H, Tandri H. Incremental value of cardiac magnetic resonance imaging in arrhythmic risk stratification of arrhythmogenic right ventricular dysplasia/cardiomyopathy-associated desmosomal mutation carriers. J Am Coll Cardiol 2013;62:1761–9.

[22] te Riele AS, James CA, Rastegar N, Bhonsale A, Murray B, Tichnell C, Judge DP, Bluemke DA, Zimmerman SL, Kamel IR, Calkins H, Tandri H. Yield of serial evaluation in at-risk family members of patients with ARVD/C. J Am Coll Cardiol 2014;64:293–301.

[23] Turrini P, Corrado D, Basso C, Nava A, Bauce B, Thiene G. Dispersion of ventricular depolarization-repolarization: a noninvasive marker for risk stratification in arrhythmogenic right ventricular cardiomyopathy. Circulation 2001;103:3075–80.

[24] Marcus FI, McKenna WJ, Sherrill D, Basso C, Bauce B, Bluemke DA, Calkins H, Corrado D, Cox MG, Daubert JP, Fontaine G, Gear K, Hauer R, Nava A, Picard MH, Protonotarios

N, Saffitz JE, Sanborn DM, Steinberg JS, Tandri H, Thiene G, Towbin JA, Tsatsopoulou A, Wichter T, Zareba W. Diagnosis of arrhythmogenic right ventricular cardiomyopathy/dysplasia: proposed modification of the task force criteria. Circulation 2010;121:1533–41.

[25] Marra MP, Leoni L, Bauce B, Corbetti F, Zorzi A, Migliore F, Silvano M, Rigato I, Tona F, Tarantini G, Cacciavillani L, Basso C, Buja G, Thiene G, Iliceto S, Corrado D. Imaging study of ventricular scar in arrhythmogenic right ventricular cardiomyopathy: comparison of 3d standard electroanatomical voltage mapping and contrast-enhanced cardiac magnetic resonance. Circ Arrhythm Electrophysiol 2012;5:91–100.

[26] Tandri H, Calkins H, Nasir K, Bomma C, Castillo E, Rutberg J, Tichnell C, Lima JA, Bluemke DA. Magnetic resonance imaging findings in patients meeting task force criteria for arrhythmogenic right ventricular dysplasia. J Cardiovasc Electrophysiol 2003;14: 476–82.

[27] Keller DI, Osswald S, Bremerich J, Bongartz G, Cron TA, Hilti P, Pfisterer ME, Buser PT. Arrhythmogenic right ventricular cardiomyopathy: diagnostic and prognostic value of the cardiac MRI in relation to arrhythmia-free survival. Int J Cardiovasc Imaging 2003;19:537–43. discussion 545–537.

[28] Schick F, Miller S, Hahn U, Nagele T, Helber U, Stauder N, Brechtel K, Claussen CD. Fat- and water-selective MR cine imaging of the human heart: assessment of right ventricular dysplasia. Invest Radiol 2000;35:311–8.

[29] Tandri H, Saranathan M, Rodriguez ER, Martinez C, Bomma C, Nasir K, Rosen B, Lima JA, Calkins H, Bluemke DA. Noninvasive detection of myocardial fibrosis in arrhythmogenic right ventricular cardiomyopathy using delayed-enhancement magnetic resonance imaging. J Am Coll Cardiol 2005;45:98–103.

[30] Silvano M, Corrado D, Kobe J, Monnig G, Basso C, Thiene G, Eckardt L. Risk stratification in arrhythmogenic right ventricular cardiomyopathy. Herzschrittmacherther Elektrophysiol 2013;24:202–8.

[31] Pinamonti B, Dragos AM, Pyxaras SA, Merlo M, Pivetta A, Barbati G, Di Lenarda A, Morgera T, Mestroni L, Sinagra G. Prognostic predictors in arrhythmogenic right ventricular cardiomyopathy: results from a 10-year registry. Eur Heart J 2011;32:1105–13.

[32] Lemola K, Brunckhorst C, Helfenstein U, Oechslin E, Jenni R, Duru F. Predictors of adverse outcome in patients with arrhythmogenic right ventricular dysplasia/cardiomyopathy: long term experience of a tertiary care centre. Heart 2005;91:1167–72.

[33] Nava A, Bauce B, Basso C, Muriago M, Rampazzo A, Villanova C, Daliento L, Buja G, Corrado D, Danieli GA, Thiene G. Clinical profile and long-term follow-up of 37 families with arrhythmogenic right ventricular cardiomyopathy. J Am Coll Cardiol 2000;36: 2226–33.

[34] Paetsch I, Reith S, Gassler N, Jahnke C. Isolated arrhythmogenic left ventricular cardiomyopathy identified by cardiac magnetic resonance imaging. Eur Heart J 2011;32:2840.

[35] Pinamonti B, Pagnan L, Bussani R, Ricci C, Silvestri F, Camerini F. Right ventricular dysplasia with biventricular involvement. Circulation 1998;98:1943–5.

[36] Deac M, Alpendurada F, Fanaie F, Vimal R, Carpenter JP, Dawson A, Miller C, Roussin I, di Pietro E, Ismail TF, Roughton M, Wong J, Dawson D, Till JA, Sheppard MN, Mohiaddin RH, Kilner PJ, Pennell DJ, Prasad SK. Prognostic value of cardiovascular magnetic resonance in patients with suspected arrhythmogenic right ventricular cardiomyopathy. Int J Cardiol 2013;168:3514–21.

[37] Aquaro GD, Pingitore A, Strata E, Di Bella G, Molinaro S, Lombardi M. Cardiac magnetic resonance predicts outcome in patients with premature ventricular complexes of left bundle branch block morphology. J Am Coll Cardiol 2010;56:1235–43.

[38] Protonotarios N, Anastasakis A, Antoniades L, Chlouverakis G, Syrris P, Basso C, Asimaki A, Theopistou A, Stefanadis C, Thiene G, McKenna WJ, Tsatsopoulou A. Arrhythmogenic right ventricular cardiomyopathy/dysplasia on the basis of the revised diagnostic criteria in affected families with desmosomal mutations. Eur Heart J 2011;32: 1097–104.

[39] Riley MP, Zado E, Bala R, Callans DJ, Cooper J, Dixit S, Garcia F, Gerstenfeld EP, Hutchinson MD, Lin D, Patel V, Verdino R, Marchlinski FE. Lack of uniform progression of endocardial scar in patients with arrhythmogenic right ventricular dysplasia/ cardiomyopathy and ventricular tachycardia. Circ Arrhythm Electrophysiol 2010;3: 332–8.
[40] Conen D, Osswald S, Cron TA, Linka A, Bremerich J, Keller DI, Pfisterer ME, Buser PT. Value of repeated cardiac magnetic resonance imaging in patients with suspected arrhythmogenic right ventricular cardiomyopathy. J Cardiovasc Magn Reson 2006;8: 361–6.

11

Echocardiographic Applications in the Diagnosis and Management of Patients with ARVC

Thomas P. Mast, Arco J. Teske,
Pieter A. Doevendans, Maarten J. Cramer

Department of Cardiology, University Medical Center Utrecht,
Utrecht, The Netherlands

INTRODUCTION

Arrhythmogenic right ventricular cardiomyopathy dysplasia (ARVC/D) is a genetically determined myocardial disease predominantly affecting the right ventricle (RV) [1]. In more than half of the ARVC patients genetic mutations are found in genes encoding for desmosomal proteins [2,3]. In North America and Europe, desmosomal mutations are predominantly found in the plakophilin-2 (*PKP2*) gene [3,4]. Desmosomal dysfunction will eventually lead to fibrofatty substitution of the myocardium. A typical distribution of disease expression is observed in the so-called triangle of dysplasia [1,5]. As a consequence of the fibrofatty substitution ventricular arrhythmias and ventricular failure occur. Recent studies show that electrical abnormalities and ventricular arrhythmias appear before extensive structural alterations [6,7].

There is no single gold standard for the diagnosis of ARVC/D and the diagnostic process is still considered challenging [8]. The diagnosis of ARVC/D is currently made on the consensus based revised 2010 Task Force criteria (TFC). Imaging criteria as part of the 2010 TFC underlines the importance of cardiac imaging in ARVC/D [9]. Cardiac magnetic resonance imaging (MRI) has a leading role in diagnosing this disease ARVC/D [10]. However, echocardiography remains a first-line diagnostic tool and plays a role in the serial evaluation of ARVC/D patients [11,12]. For decades, echocardiography has proven to be a valuable tool in studying ARVC/D [12,13]. Recent developments in RV echocardiography are changing the current and future role of echocardiography in ARVC/D.

ROLE OF ECHOCARDIOGRAPHY IN DIAGNOSIS OF ARVC/D

Echocardiography is a noninvasive, relatively inexpensive, and widely available imaging tool to evaluate patients with suspected ARVC/D. ARVC/D is diagnosed according to major and minor criteria stated in the 2010 TFC [9]. The 2010 TFC includes different categories: (1) structural RV alterations, (2) tissue characterization, (3) depolarization and repolarization abnormalities, (4) ventricular arrhythmias, and (5) family history and genetic data [9]. The presence of two major criteria, one major and two minor criteria, or four minor criteria from different categories indicates a definite diagnosis of ARVC/D. RV structural alterations are visualized using different imaging modalities including cardiac MRI, angiography, and echocardiography. Structural alterations consists of regional RV wall motion abnormalities in combination with RV dilatation or global RV systolic dysfunction. Modality-specific values concerning global RV dysfunction and RV dilatation are provided in the 2010 TFC (Table 11.1). Normal values for echocardiographic TFC are chosen for high specificity (76–95%), resulting in reduced sensitivity (55–87%) [9].

Regional Wall Motion Abnormalities

Detection of a regional RV wall motion abnormality is required to score a major or minor criterion regardless of outflow tract dilatation or systolic dysfunction [9]. While there seem to be clear predilection sites within the RV, these regional wall motion abnormalities can occur in any part of the RV. Therefore, optimal visual RV wall motion analysis by dedicated echocardiographic views is crucial and all RV walls should be visualized during the examination. Accurate assessment of the RV is generally considered difficult due to its complex crescent shape and retrosternal anatomical position. Contrast-enhanced echocardiography can be used to improve delineation of the RV and enables more accurate RV wall motion analyses if there is suboptimal imaging quality [14]. However, assessment of RV regional wall motion is often possible without contrast by using multiple echocardiographic views (Fig. 11.1). While standard echocardiographic examinations are predominantly focused on the left ventricle (LV), the average echo lab might be unfamiliar with a complete RV examination. Standard RV views that should be part of an echocardiographic ARVC/D protocol are depicted in Fig. 11.1. For both MRI and echocardiography, RV imaging remains challenging [15,16].

TABLE 11.1   2010 Task Force Criteria. Global or Regional Dysfunction and Structural Alterations

Echocardiography	Sensitivity (%)	Specificity (%)
• Regional RV akinesia, dyskinesia, or aneurysm and one of the following parameters		
Major		
PLAX RVOT ≥ 32 mm or corrected for BSA ≥ 19 mm/m^2 (end diastolic)	75	95
PSAX RVOT ≥ 36 mm or corrected for BSA ≥ 21 mm/m^2 (end diastolic)	62	95
RV-fractional area change ≤ 33%	55	95
Minor		
PLAX RVOT ≥ 29 to <32 mm or corrected for BSA ≥ 16 to <19 mm/m^2 (end diastolic)	87	87
PSAX RVOT ≥ 32 to <36 mm or corrected for BSA ≥ 18 to <21 mm/m^2 (end diastolic)	80	80
RV-fractional area change >33% to ≤40%	76	76
MRI	**Sensitivity (%)**	**Specificity (%)**
• Regional RV akinesia, dyskinesia, or dyssynchronous RV contraction	Male/female	Male/female
Major		
Ratio of RV end-diastolic volume to BSA ≥ 110 mL/m^2 (male) or ≥100 mL/m^2 (female) RV-ejection fraction ≤ 40%	76/68	90/98
Minor		
Ratio of RV end-diastolic volume to BSA ≥100 to <110 mL/m^2 (male) or ≥90 to <100 mL/m^2 (female) RV ejection fraction >40% to ≤45%	79/89	85/97
RV angiography		
Major		
Regional RV akinesia, dyskinesia, or aneurysm		
Minor		
No minor criterion is specified for this modality.		

RV, right ventricle; PLAX/PSAX, parasternal long/shorts axis; BSA, body surface area; mm, millimeter; mL, milliliter; m, meter; MRI, magnetic resonance imaging.

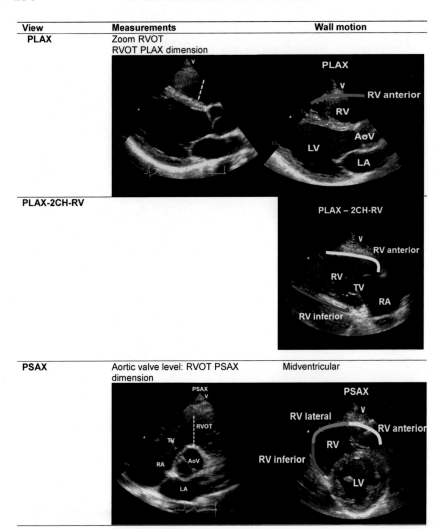

FIGURE 11.1 **RV echocardiographic views.** Views are in addition to standard conventional echocardiography as provided by the American Society of Echocardiography (ASE). BSA, body surface area; PLAX/PSAX, parasternal long/short axis view; 2CH, two-chamber view; 4CH, four-chamber view; RV, right ventricular; LV, left ventricle; LA, left atrium; RA, right atrium; RVOT, right ventricular outflow tract; RV-FAC, right ventricular-fractional area change; TAPSE, tricuspid annular plane systolic excursion.

Ideally, both modalities should be performed in patients suspected of ARVC/D, especially in individuals with poor echocardiographic images or with borderline findings. Therefore, visualizing all RV walls during echocardiography is of utmost importance for comparison between the two imaging modalities. Comparing results of RV wall motion analyses by MRI and echocardiography may improve the accuracy of interpretation.

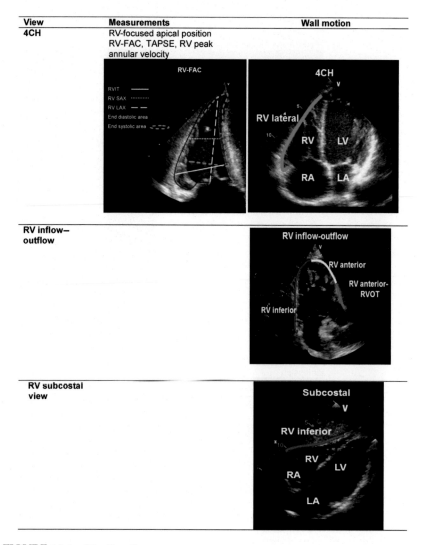

View	Measurements	Wall motion
4CH	RV-focused apical position RV-FAC, TAPSE, RV peak annular velocity	
RV inflow–outflow		
RV subcostal view		

FIGURE 11.1   *(Continued).*

Akinetic, dyskinetic, or aneurysmal RV wall segments are considered abnormal in the 2010 TFC. An akinetic segment shows no movement or is moved passively by adjacent segments. Dyskinesia is defined as a segment moving outward during systole. A localized outward bulging of the ventricular wall is described as an aneurysm. All mentioned regional wall motion abnormalities are frequently seen in ARVC/D [11]. Distribution of regional wall motion abnormalities as detected by echocardiography is provided in Table 11.2 [17]. Hypokinesia is no longer part of the 2010 TFC. Indeed, this phenomenon, while abnormal when observed in the LV,

TABLE 11.2   Distribution of RV Wall Motion Abnormalities in ARVC Patients

Yoerger et al. [11] N = 29 probands		Teske et al. (2015, unpublished data)	
RV regional WMA (%)	79		
RVOT (%)	45	RVOT (%)	60
Anteroseptal (%)	55	Anterior (%)	43
Anterior (%)	70	Inferior (%)	40
Apex (%)	72	Lateral (%)	90
Septal (%)	55	Apex (%)	50
Inferior basal (%)	59		
Inferior apical (%)	52		

RV, right ventricle; RVOT, right ventricular outflow tract.

seems to be a normal variant in the RV since it is frequently observed in healthy individuals, particularly at the insertion point of the moderator band [17].

Right Ventricular Dimensions

Regional and global RV dilatation are well-described features in ARVC/D (Fig. 11.2) [11]. In contrast to MRI, segmental dilation of the right ventricular outflow tract (RVOT) was chosen for the echocardiographic 2010 TFC rather than global RV dilatation (Table 11.1 and Fig. 11.1) [9]. Simple geometrical assumptions, as frequently used to assess LV dimensions are invalid in the RV and prevent optimal volumetric calculation based on 2D acquisition. Normal values for the RVOT are provided both corrected and uncorrected for body surface area (BSA) (Table 11.1) [9]. Accurate measurement of RVOT dimensions can be challenging due to an oblique plane resulting in overestimation of RVOT dimensions. Moreover, a clear delineation of the RV anterior endocardium may be difficult. Of note, current RV echocardiographic guidelines are not in concordance with the cutoff values for RVOT dilatation in the 2010 TFC [18]. Also, dilatation of the RVOT is a nonspecific finding since it is frequently seen in individuals with chronic volume overload (e.g., shunting, endurance athletes) or pressure overload.

RV Systolic Function

As previously mentioned, RV optimal volumetric calculation based on 2D acquisition is difficult. However, several established parameters

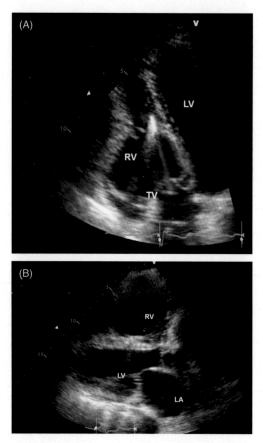

FIGURE 11.2   **Echocardiographic examples of ARVC/D patients.** (A) Echocardiographic image: modified apical four-chamber RV view. The patient was diagnosed with ARVC according to 2010 TFC. An aneurysm is seen in the subtricuspid area. The reverberating structure is an ICD electrode. (B) Echocardiographic image: parasternal long axis. The patient was diagnosed with ARVC/D according to 2010 TFC. The RV is severely dilated. The RVOT was measured 70 mm. RV, right ventricle; LV, left ventricle; LA, left atrium; ARVC/D, arrhythmogenic right ventricular cardiomyopathy dysplasia; TFC, Task Force criteria; ICD, implantable cardioverter device; RVOT, right ventricular outflow tract; mm, millimeter.

for RV systolic function are known. Tricuspid annular plane systolic excursion (TAPSE), peak systolic RV annular velocity, and RV-fractional area change (RV-FAC) provide quick evaluation of global RV function and have been found to be decreased in ARVC/D [11,19].

RV-FAC was chosen as an indicator of RV systolic dysfunction in the 2010 echocardiographic TFC [9]. In the apical four-chamber view, it reflects the area change of the RV during systole [18]. RV-FAC is a single plane measurement and correlates with CMR-derived RV ejection fraction [20]. Clear delineation of the endocardium may be challenging due to the trabecularization of the RV, and off angle recordings typically underestimate the actual RV size. This should be taken into account when interpreting these findings; since it can lead to false-negative results (overestimating RV systolic function).

TAPSE and peak systolic RV annular velocity are not hampered by these limitations and provide insight into RV systolic function if RV-FAC cannot be accurately acquired. Also, they show a high inter- and intra-observer variability and are well-established robust parameters for global RV systolic function. Unfortunately, for both TAPSE and peak systolic RV annular velocity, values are not provided in the TFC for diagnosing ARVC/D.

Both TAPSE and RV-FAC, are indicators of RV systolic dysfunction, and contain prognostic information. Reduced values of TAPSE and RV-FAC have been correlated with adverse outcome in an ARVC/D cohort [12].

Left Ventricular Involvement

In ARVC/D, the typically involved areas are (1) the subtricuspid region, (2) RVOT, and (3) LV posterolateral wall [5]. Traditionally, the RV apex has been seen as an important region that was involved early in this disease and included in the so-called "triangle of dysplasia" [9]. However, recent studies have clearly shown that the apex is involved in advanced stages of disease [5,21,22]. Frequent left-sided involvement has resulted in a "displacement" of the classical triangle (Fig. 11.3) [5].

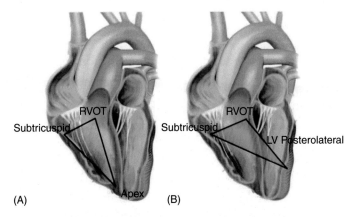

FIGURE 11.3 **Triangle of dysplasia.** (A) Original triangle of dysplasia consisting of the following regions: subtricuspid region, RV outflow tract (RVOT), and apical region. (B) Triangle of dysplasia displaced. Nowadays, the left ventricular posterolateral wall seems to be a predilection site of ARVC/D and is more commonly involved compared to the apical region.

LV involvement is also associated with the prognosis of ARVC [23,24]. LV ejection fraction, was repeatedly found to be an independent predictor for adverse outcome [23,24]. For the serial evaluation of an ARVC/D patient, assessing LV function by Simpson's biplane measurement of LV ejection fraction provides insight into the individual course of the disease.

FUTURE DEVELOPMENTS IN ECHOCARDIOGRAPHY

Three-Dimensional RV Echocardiography

2D measurements on echocardiography are likely inferior to the 3D volumetric analysis performed on MRI. Therefore, 3D RV echocardiography has been evaluated to overcome the limitations of 2D RV echocardiography (Fig. 11.4) [25,26]. 3D RV echocardiography seems to be comparable to MRI-derived RV volumes and RV ejection fraction. Commercial software packages are readily available [25]. However, 3D echocardiography can be limited by acoustic dropout of the RV. This limitation may be most pronounced in dilated RV chambers as is common in advanced ARVC/D. Nonetheless, several studies showed good agreement in 3D calculated volumes between MRI and echocardiography in ARVC [25,26].

Another approach of 3D RV echocardiography is the use of knowledge-based 3D reconstruction of the RV (Fig. 11.4) [27,28]. With the use of anatomical landmarks acquired on 2D images from different views, the algorithm uses the distances between RV landmarks to match the landmarks to a catalog of subjects with similar RV anatomy. This approach overcomes the limitation of acoustic dropout and the necessity of perfect image quality of the entire RV endocardium from one single view. Further validation is necessary and the feasibility of 3D reconstruction in ARVC/D patients is currently unknown. Nevertheless, 3D RV echocardiography is promising for further validation and may be implemented in the ARVC/D diagnostic process in the near future.

Tissue Deformation Imaging

Echocardiographic tissue deformation imaging is a relative new technique providing quantitative analysis of regional wall motion [29]. Currently, two different techniques are commercially available: tissue Doppler imaging and speckle tracking. Both techniques rely on different principles to calculate both regional as well as global deformation and deformation rate [29]. Both are feasible in the RV lateral free wall and the

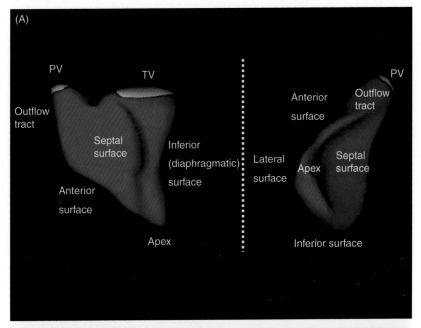

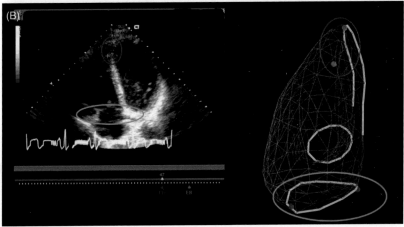

FIGURE 11.4   **Examples of three-dimensional RV imaging.** (A) 3D RV echocardiographic imaging providing insight into RV anatomy. (B) 3D acquisition with knowledge-based reconstruction. Multiple landmarks, chosen in 2D images (left), are used to reconstruct the RV anatomy (right). PV, pulmonary valve; TV, tricuspid valve.

calculated values are generally comparable in ARVC/D patients [19,30]. Optimal deformation analysis of the RV anterior, inferior, and outflow tract is not suitable due to limitations in temporal resolution and angle dependency.

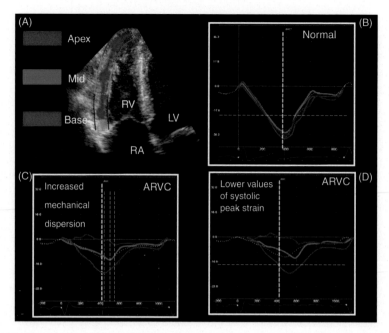

FIGURE 11.5 **Tissue deformation imaging.** (A) Tracing of the RV lateral free wall divided into the basal, mid, and apical segment. (B) Normal deformation curve. Horizontal broken line indicates normal peak systolic strain. (D) Deformation curve of ARVC/D patient. Basal (subtricuspid) segment (red) and mid segment (blue) are showing lower values of systolic peak strain. (C) Vertical broken line indicates timing of peak strain. Mechanical dispersion is higher compared to normal deformation curve indicating mechanical dyssynchrony. ARVC/D, arrhythmogenic right ventricular cardiomyopathy dysplasia; RV, right ventricle; LV, left ventricle; RA, right atrium.

A major advantage of tissue deformation imaging compared to conventional echocardiographic wall motion analysis is the ability to quantify regional wall motion rather than qualitative assessment as stated in the 2010 TFC. Quantifying wall motion facilitates the detection of subtle regional myocardial dysfunction. Secondly, deformation imaging analysis is not hampered by misinterpretation of wall motion caused by tissue tethering. A third advantage is the ability to compare different myocardial segments with time to measure mechanical synchrony.

Peak systolic deformation or peak strain appears to be a reliable parameter to differentiate between normal and abnormal RV segments in ARVC/D (Fig. 11.5) [19,30–32]. The diagnostic value of peak systolic strain is better than conventional echocardiographic parameters and sensitivity and specificity are both reported to be greater than 90% [19,30–33].

Besides regional and global quantitative wall motion analysis tissue deformation imaging provides insight into mechanical synchrony [34]. Due to the electrical–mechanical coupling, electrical disturbances can be studied by mechanical activation patterns [34]. Electrical abnormalities may precede structural alteration and can play an important role in early diagnosis in the absence of global RV dysfunction [7]. Mechanical dispersion of the timing of peak systolic deformation has been shown to be associated with arrhythmic outcome in ARVC/D [34]. The results of tissue deformation imaging are promising. Further validation is needed to determine the role of these tools in ARVC/D.

ROLE OF ECHOCARDIOGRAPHY IN THE ERA OF MRI

In a direct comparison between these imaging modalities, MRI has been found to have superior diagnostic value compared to conventional echocardiography [15]. Nevertheless, MRI is prone to false-positive findings due to an overreliance on detection of intramyocardial fat and wall thinning [16]. In addition, evaluation of RV wall motion in the area of the moderator band insertions remains cumbersome by MRI [35]. Therefore, both modalities should be used in the diagnostic process to prevent false-positive and false-negative results of either modality. MRI and echocardiography should be considered complementary diagnostic abnormalities.

In ARVC/D patients, implantable cardioverter defibrillators (ICDs) are frequently inserted to prevent sudden arrhythmic death in addition to medical therapy. Unfortunately, these metallic devices consequently make MRI unsuitable for repeated evaluation of ARVC/D patients. For the purpose of monitoring ARVC/D patients, echocardiography is the current preferred imaging technology. Assessing RV and LV systolic function is easily obtainable by 2D echocardigraphy and provides excellent prognostic value in ARVC/D patients [23,24,34].

Emerging new echocardiographic techniques, especially 3D RV echocardiography and tissue deformation imaging, may increase the performance of echocardiography, overcoming the shortcomings of conventional RV echocardiography.

References

[1] Marcus FI, Fontaine GH, Guiraudon G, Frank R, Laurenceau JL, Malergue C, Grosgogeat Y. Right ventricular dysplasia: a report of 24 adult cases. Circulation 1982;65:384–98.
[2] Sen-Chowdhry S, Syrris P, McKenna WJ. Role of genetic analysis in the management of patients with arrhythmogenic right ventricular dysplasia/cardiomyopathy. J Am Coll Cardiol 2007;50:1813–21.
[3] Cox MG, van der Zwaag PA, van der Werf C, et al. Arrhythmogenic right ventricular dysplasia/cardiomyopathy: pathogenic desmosome mutations in index-patients predict outcome of family screening: Dutch arrhythmogenic right ventricular dysplasia/cardiomyopathy genotype-phenotype follow-up study. Circulation 2011;123:2690–700.
[4] Fressart V, Duthoit G, Donal E, et al. Desmosomal gene analysis in arrhythmogenic right ventricular dysplasia/cardiomyopathy: spectrum of mutations and clinical impact in practice. Europace 2010;12:861–8.

[5] Te Riele AS, James CA, Philips B, et al. Mutation-positive arrhythmogenic right ventricular dysplasia/cardiomyopathy: the triangle of dysplasia displaced. J Cardiovasc Electrophysiol 2013;24:1311–20.

[6] Cerrone M, Noorman M, Lin X, Chkourko H, Liang FX, van der Nagel R, Hund T, Birchmeier W, Mohler P, van Veen TA, van Rijen HV, Delmar M. Sodium current deficit and arrhythmogenesis in a murine model of plakophilin-2 haploinsufficiency. Cardiovasc Res 2012;95:460–8.

[7] te Riele AS, James CA, Rastegar N, Bhonsale A, Murray B, Tichnell C, Judge DP, Bluemke DA, Zimmerman SL, Kamel IR, Calkins H, Tandri H. Yield of serial evaluation in at-risk family members of patients with ARVD/C. J Am Coll Cardiol 2014; 64:293–301.

[8] Basso C, Corrado D, Marcus FI, Nava A, Thiene G. Arrhythmogenic right ventricular cardiomyopathy. Lancet 2009;373:1289–300.

[9] Marcus FI, McKenna WJ, Sherrill D, et al. Diagnosis of arrhythmogenic right ventricular cardiomyopathy/dysplasia: proposed modification of the task force criteria. Circulation 2010;121:1533–41.

[10] te Riele AS, Tandri H, Bluemke DA. Arrhythmogenic right ventricular cardiomyopathy (ARVC): cardiovascular magnetic resonance update. J Cardiovasc Magn Reson 2014;16:50.

[11] Yoerger DM, Marcus F, Sherrill D, Calkins H, Towbin JA, Zareba W, Picard MH. Echocardiographic findings in patients meeting task force criteria for arrhythmogenic right ventricular dysplasia: new insights from the multidisciplinary study of right ventricular dysplasia. J Am Coll Cardiol 2005;45:860–5.

[12] Saguner AM, Vecchiati A, Baldinger SH, et al. Different prognostic value of functional right ventricular parameters in arrhythmogenic right ventricular cardiomyopathy/dysplasia. Circ Cardiovasc Imaging 2014;7(2):230–9.

[13] Blomstrom-Lundqvist C, Beckman-Suurkula M, Wallentin I, Jonsson R, Olsson SB. Ventricular dimensions and wall motion assessed by echocardiography in patients with arrhythmogenic right ventricular dysplasia. Eur Heart J 1988;9:1291–302.

[14] Tosoratti E, Badano LP, Gianfagna P, Baldassi M, Proclemer A, Capelli C, Fioretti PM. Improved delineation of morphological features of arrhythmogenic right ventricular cardiomyopathy with the use of contrast-enhanced echocardiography. J Cardiovasc Med (Hagerstown) 2006;7:566–8.

[15] Borgquist R, Haugaa KH, Gilljam T, Bundgaard H, Hansen J, Eschen O, Jensen HK, Holst AG, Edvardsen T, Svendsen JH, Platonov PG. The diagnostic performance of imaging methods in ARVC using the 2010 Task Force criteria. Eur Heart J Cardiovasc Imaging 2014;15:1219–25.

[16] Bomma C, Rutberg J, Tandri H, Nasir K, Roguin A, Tichnell C, Rodriguez R, James C, Kasper E, Spevak P, Bluemke DA, Calkins H. Misdiagnosis of arrhythmogenic right ventricular dysplasia/cardiomyopathy. J Cardiovasc Electrophysiol 2004;15:300–6.

[17] Teske AJ, Cox MG, Te Riele AS, De Boeck BW, Doevendans PA, Hauer RN, Cramer MJ. Early detection of regional functional abnormalities in asymptomatic ARVD/C gene carriers. J Am Soc Echocardiogr 2012;25:997–1006.

[18] Rudski LG, Lai WW, Afilalo J, Hua L, Handschumacher MD, Chandrasekaran K, Solomon SD, Louie EK, Schiller NB. Guidelines for the echocardiographic assessment of the right heart in adults: a report from the American Society of Echocardiography endorsed by the European Association of Echocardiography, a registered branch of the European Society of Cardiology, and the Canadian Society of Echocardiography. J Am Soc Echocardiog 2010;23:685–713.

[19] Teske AJ, Cox MG, De Boeck BW, Doevendans PA, Hauer RN, Cramer MJ. Echocardiographic tissue deformation imaging quantifies abnormal regional right ventricular function in arrhythmogenic right ventricular dysplasia/cardiomyopathy. J Am Soc Echocardiogr 2009;22:920–7.

[20] Anavekar NS, Gerson D, Skali H, Kwong RY, Yucel EK, Solomon SD. Two-dimensional assessment of right ventricular function: an echocardiographic-MRI correlative study. Echocardiography 2007;24:452–6.

[21] Lindstrom L, Nylander E, Larsson H, Wranne B. Left ventricular involvement in arrhythmogenic right ventricular cardiomyopathy – a scintigraphic and echocardiographic study. Clin Physiol Funct Imaging 2005;25:171–7.

[22] Sen-Chowdhry S, Syrris P, Ward D, Asimaki A, Sevdalis E, McKenna WJ. Clinical and genetic characterization of families with arrhythmogenic right ventricular dysplasia/cardiomyopathy provides novel insights into patterns of disease expression. Circulation 2007;115:1710–20.

[23] Pinamonti B, Dragos AM, Pyxaras SA, Merlo M, Pivetta A, Barbati G, Di Lenarda A, Morgera T, Mestroni L, Sinagra G. Prognostic predictors in arrhythmogenic right ventricular cardiomyopathy: results from a 10-year registry. Eur Heart J 2011;32:1105–13.

[24] Mast, et al. Left ventricular involvement in arrhythmogenic right ventricular dysplasia/cardiomyopathy assessed by echocardiography predicts adverse clinical outcome. J Am Soc Echocardiogr 2015.

[25] Prakasa KR, Dalal D, Wang J, et al. Feasibility and variability of three dimensional echocardiography in arrhythmogenic right ventricular dysplasia/cardiomyopathy. Am J Cardiol 2006;97:703–9.

[26] Kjaergaard J, Hastrup Svendsen J, Sogaard P, Chen X, Bay Nielsen H, Kober L, Kjaer A, Hassager C. Advanced quantitative echocardiography in arrhythmogenic right ventricular cardiomyopathy. J Am Soc Echocardiogr 2007;20:27–35.

[27] Bhave NM, Patel AR, Weinert L, Yamat M, Freed BH, Mor-Avi V, Gomberg-Maitland M, Lang RM. Three-dimensional modeling of the right ventricle from two-dimensional transthoracic echocardiographic images: utility of knowledge-based reconstruction in pulmonary arterial hypertension. J Am Soc Echocardiogr 2013;26:860–7.

[28] Dragulescu A, Grosse-Wortmann L, Fackoury C, Mertens L. Echocardiographic assessment of right ventricular volumes: a comparison of different techniques in children after surgical repair of tetralogy of Fallot. Eur Heart J Cardiovasc Imaging 2012;13:596–604.

[29] Teske AJ, De Boeck BW, Melman PG, Sieswerda GT, Doevendans PA, Cramer MJ. Echocardiographic quantification of myocardial function using tissue deformation imaging, a guide to image acquisition and analysis using tissue Doppler and speckle tracking. Cardiovasc Ultrasound 2007;5:27.

[30] Prakasa KR, Wang J, Tandri H, et al. Utility of tissue Doppler and strain echocardiography in arrhythmogenic right ventricular dysplasia/cardiomyopathy. Am J Cardiol 2007;100:507–12.

[31] Kjaergaard J. Assessment of right ventricular systolic function by tissue Doppler echocardiography. Dan Med J 2012;59:B4409.

[32] Aneq MA, Engvall J, Brudin L, Nylander E. Evaluation of right and left ventricular function using speckle tracking echocardiography in patients with arrhythmogenic right ventricular cardiomyopathy and their first degree relatives. Cardiovasc Ultrasound 2012;10:37.

[33] Vitarelli A, Cortes Morichetti M, Capotosto L, De Cicco V, Ricci S, Caranci F, Vitarelli M. Utility of strain echocardiography at rest and after stress testing in arrhythmogenic right ventricular dysplasia. Am J Cardiol 2013;111:1344–50.

[34] Sarvari SI, Haugaa KH, Anfinsen OG, Leren TP, Smiseth OA, Kongsgaard E, Amlie JP, Edvardsen T. Right ventricular mechanical dispersion is related to malignant arrhythmias: a study of patients with arrhythmogenic right ventricular cardiomyopathy and subclinical right ventricular dysfunction. Eur Heart J 2011;32:1089–96.

[35] Sievers B, Addo M, Franken U, Trappe HJ. Right ventricular wall motion abnormalities found in healthy subjects by cardiovascular magnetic resonance imaging and characterized with a new segmental model. J Cardiovasc Magn Reson 2004;6:601–8.

Other Imaging Modalities in the Evaluation of Patients with ARVC/D

Aiden Abidov, Ahmed K. Pasha,
Isabel B. Oliva

Department of Medicine/Division of Cardiology and Department
of Medical Imaging, University of Arizona, Tucson, AZ, USA

INTRODUCTION

While echocardiography and cardiac MR are the main imaging modalities utilized for the diagnosis of ARVC/D, there are other imaging modalities for the diagnostic evaluation or use as a complementary diagnostic tool [1]. These include cardiac CTA (CCTA) as well as several nuclear imaging methods (Table 12.1). They are used to evaluate the extent severity of fibrosis, quantitate fibro-fatty infiltration, involvement of the left ventricle, and extent of the arrhythmogenic substrate. They can perform follow-up imaging in patients with implanted ICD, and assess great vessels and extracardiac thoracic structures. Furthermore, some novel applications are useful in understanding the arrhythmic substrate and risk of sudden death. Other advanced techniques, such as adenoreceptor density measurement, cardiac dyssynchrony, phase analysis, and 3D-based volumetric and speckle tracking data of the left and the right heart, are being introduced into clinical practice.

Cardiac MRI in the Diagnosis, Clinical Management and Prognosis of Arrhythmogenic Right Ventricular Cardiomyopathy/Dysplasia

TABLE 12.1 Cardiac Imaging Modalities Utilized in Patients With Known or Suspected ARVC/D

Imaging modality	Clinical target	Potential benefits	Associated risks/concerns
Cardiac CTA	Assessment of coronary arteries	Noninvasive diagnosis of CAD, especially when ischemic RV changes are suspected	Radiation exposure; recent developments allow achievement of ultralow radiation dose
	Assessment of pulmonary arteries and other great vessels	Exclusion of thromboembolic disease mimicking ARVC/D; assessment of possible congenital abnormalities associated with an abnormal RV	Iodine contrast exposure can cause renal dysfunction/contrast-induced nephropathy and a significant risk in patients with severe contrast allergy/anaphylaxis
	LV and RV volumes, global and regional function	Primary or confirmatory assessment in patients with suboptimal MRI	Suboptimal imaging in patients with highly irregular and/or fast heart rate
	RV wall thinning	Improved delineation of the arrhythmogenic substrate	
	Quantitation of the RV and LV fatty infiltration	Capability to accurately quantify myocardial fat in the RV and LV	Requires evaluation of the imaging data on an independent 3D workstation for maximal diagnostic yield
	Primary imaging modality in patients with contraindications to MRI, or suboptimal MRI imaging	3D acquisition permits fast acquisition and complete assessment of the RV and LV in unstable patients, "MRI – challenging" cases	May demonstrate some degree of variability in comparison with the RV and LV volumes and function acquired with other modalities (Cardiac MRI or Echo)
	Follow up imaging in patients with permanent pacemaker/ICD and/or suboptimal RV visualization on MRI	Diagnostic image quality and excellent inter- and intraobserver agreement in serial studies	

Gated blood pool single photon emission computed tomography (SPECT) and equilibrium radionuclide angiocardiography (ERNA)	Assessment of extracardiac structures (lungs, mediastinum)	Extracardiac pathology is a suspected cause of RV abnormality (sarcoidosis, lung disorders with cor pulmonale leading to RV enlargement/failure, etc)	Consider all potential risks and benefits as well as alternative options while choosing CTA for young and female patients (due to the radiation exposure risk)
	LV and RV volumes, global systolic function	Primary or confirmatory assessment in patients with suboptimal MRI Excellent repeatability and reproducibility in serial studies Multiple software tools available for quantitation of the LV and RV volumes	Low spatial resolution (difficult to definitively assess regional wall motion abnormalities) Radiation exposure
MIBG imaging	Evaluation of myocardial adenoreceptor density	Added value in identification of patients at risk of sudden cardiac death in addition to the volume, structure and function	Radiation exposure Limited clinical experience Limited availability of 3D data/MIBG SPECT and automated algorithms
Cardiac F18-FDG positron emission tomography (PET)	Evaluation of inflammation (in cases of sarcoidosis, myocarditis and other myocardial diseases mimicking ARVC/D)	Enhanced diagnostic accuracy in distinguishing inflammatory myocardial response	Radiation exposure May require patient preparation (depending on the blood glucose levels)

Appropriate choice of the imaging modality can significantly shorten the time-to-diagnosis, and therapy, avoid unnecessary testing, and decrease the number of nondiagnostic tests. These factors may result in substantial cost savings as well as selection of the most effective management strategy. Another important aspect is consideration of the multimodality imaging approach for the diagnosis and management of these patients.

In this chapter, we describe novel applications of advanced cardiac imaging modalities in patients with known or suspected ARVC/D and provide an approach to utilization of these methods for specific clinical situations.

Advanced Echo and MRI Applications

There are numerous benefits of echocardiography in patients with known or suspected ARVC/D. Unfortunately, currently available IV-contrast agents used in echocardiography, do not provide adequate RV contrast and thus, are not useful in enhancing our ability to diagnose global or regional RV dysfunction. Frequently, volumetric evaluation of the RV and quantification of RV function by regular 2D-echo can be a challenge in many patients with known or suspected ARVC/D. Utilization of modern technology with full volume acquisition of the RV can provide a very detailed and anatomically correct 3D model of the RV, thus facilitating measurements of RV volumes and RVEF (Fig. 12.1) [2]. The structural and functional assessment of the RV constitutes an integral part of the diagnostic criteria. 3D echocardiography is a relatively new technique that is able to accurately evaluate the structure and function of the RV as compared to 2D echocardiography [3]. Prakasa et al. [4], showed that the RV volumes and RVEF calculated using 3D echocardiography are comparable to that obtained from CMR. 3D echocardiography can be used as a modality of choice for the assessment and evaluation of progression of ARVC/D in patients who usually have defibrillators that preclude the use of CMR [5].

Recently, there have been several preliminary reports describing the utility of 2D speckle tracing echocardiography (STE) for RV functional assessment in patients with known or suspected ARVC/D. This methodology takes into account strain, strain rate, displacement, and velocity; hence accurately assessing RV mechanics [6]. Based on these reports, RV volume and contractility data provide important diagnostic and prognostic information in these patients [7]. STE is a relatively new technique, which is highly reproducible with very low inter- and intraobserver agreement. In Chapter 11, Mast et al. provides a detailed description of the RV strain methodology for the diagnostic evaluation of patients with ARVC/D.

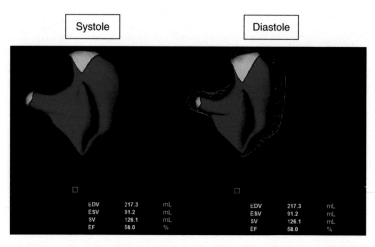

FIGURE 12.1 **Three-dimensional dynamic RV model, demonstrating automated algorithm for the RV volumetric measurements and RVEF (Tom Tec postprocessing software).**

The presence of RV pathology, such as RV akinesia, dyskinesia, dyssynchronous contraction, calculation of volumes, RVEF, and tissue characterization can be determined by the CMR protocol [1]. However, lack of standardization and increased inter- and intraobserver variability are an intrinsic problem with the qualitative, visual approach. Most recently, CMR-based LV and RV strain applications demonstrated superior diagnostic performance [8] and may have significant value in diagnosing the early phase of the disease as well as objectively defining focal RV wall motion abnormalities/myocardial deformation [9] (Fig. 12.2). Longitudinal and circumferential strain and strain rates (rate of change of muscle deformation) are significantly decreased in patients with ARVC/D. Early in the pathophysiology of ARVC/D when the RV ejection fraction is in the normal range, global longitudinal strain rate is significantly reduced in borderline ARVC/D patients thus detecting the disease earlier than with conventional imaging [10]. There is minimal intra- or interobserver variability as there is no visual analysis of wall motion. It appears to be a useful tool to detect and quantify ARVC/D but prospective longitudinal studies are required to establish its role.

Exercise Stress Testing

Exercise plays a role in unmasking ARVC/D to elicit ventricular arrhythmias that can indicate a propensity to sudden cardiac death [11]. During

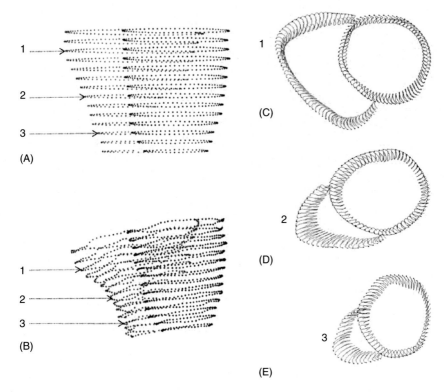

FIGURE 12.2 **Concept of the 3D myocardial displacement utilizing 3D midline motion trajectories.** (A, B) show the RV (red) and LV (blue) positions during end diastole and end systole, respectively; (C,D, and E) represent the motion trajectories for the respective slices 1, 2, and 3. *Reproduced with permission from Ref. [9].*

exercise, there is an increased preload and right-sided heart chambers are subjected to increased stretch [12]. The forces that are generated up-regulate the expression of junction proteins (Plakoglobin, N-cadherin, and Desmoplakin), which are mutated, causing structural disarray of cardiac tissue. This phenotype of ARVC/D depends upon the penetrance and component of protein affected by mutation [13].

The pathogenesis of ARVC/D includes a latent phase where there are no histological or structural changes to suggest the presence of disease or a significant risk for sudden cardiac death [14]. Exercise testing can expose the arrhythmic substrate present in asymptomatic ARVC/D gene mutation carriers. During exercise stress testing, the appearance of epsilon waves (new) and/or PVCs of superior axis LBBB pattern signify an increased risk of development of arrhythmias. Exercise stress testing can be helpful in identifying patients who are developing an

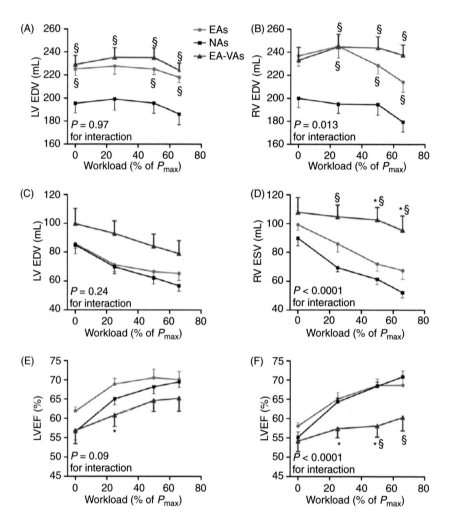

FIGURE 12.3 **CMR-derived biventricular volume changes during exercise.** Legend from the original publication: changes in EDV, end-diastolic volume; ESV, end-systolic volume; and EF, ejection fraction; during incremental exercise are shown for endurance athletes with ventricular arrhythmias (EA-VAs) (red), EAs (green), and nonathletes (NAs) (blue). *Reproduced with permission from Ref. [16].*

arrhythmic substrate that may cause fatal arrhythmias. It is also useful in surveillance and to guide exercise prescription [15].

When athletes with significant arrhythmias were compared to asymptomatic endurance athletes and normal nonathletes, they were found to have a significantly decreased RV functional reserve (Figs 12.3 and 12.4) [16].

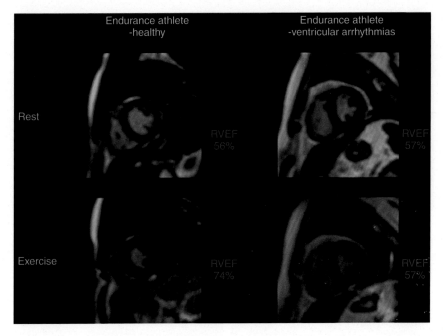

FIGURE 12.4   **Reduced stress RV function in an athlete with arrhythmias when compared with normal exercise RV function in a healthy endurance athlete.** *Reproduced with permission from Ref. [16].*

This observation suggests that stress CMR or stress echo may successfully identify patients with borderline resting anatomy and function among endurance athletes with arrhythmias or symptoms (palpitations or syncope). A substantial number of these patients may have subclinical disease that is manifest during the stress. In the future, stress testing may be considered for the TFC inclusion as one of the diagnostic criteria of early disease.

Cardiac CTA

Cardiac CTA (CCTA) is an imaging modality for assessment of patients with ARVC/D not only volumetric data and RV dimensions, but also it can provide an evaluation of tissue characterization (fibrofatty infiltration of the ventricles, see Fig. 12.5). Since CCTA is a 3D modality, it can show any view of the RV during postprocessing (Figs 12.6 and 12.7). Despite high reproducibility, excellent spatial resolution, and lesser operator dependency, there is limited experience with this modality as compared to CMR [17]. However, CCTA has a unique value in patients with prosthetic material that is not compatible with

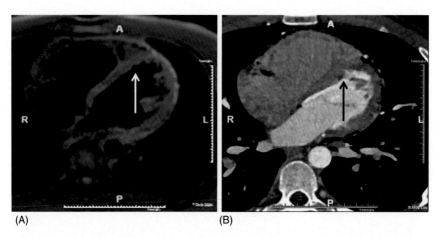

(A)   (B)

FIGURE 12.5 **Fatty infiltration of the interventricular septum (arrows) in a patient with ARVC/D.** (A) Cardiac MRI (triple inversion recovery sequence); (B) cardiac CTA (corresponding axial views).

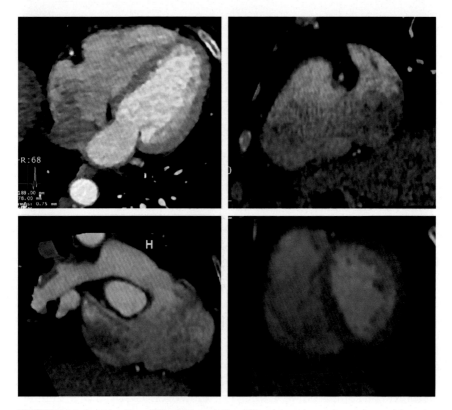

FIGURE 12.6 **Anatomy of the RV on cardiac CTA.**

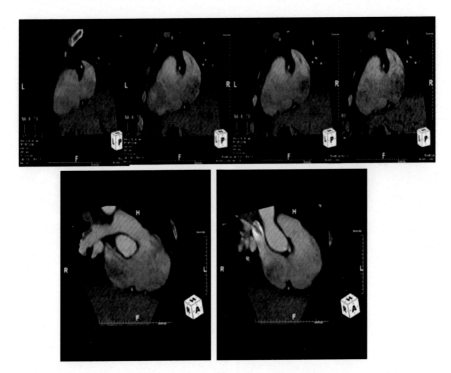

FIGURE 12.7   **CTA in assessment of the RV size and function.** 3D nature of the imaging permits the evaluation of an infinite variety of the RV views, including the "triangle of dysplasia."

MRI, such as pacemakers/implantable cardioverter-defibrillators (in centers where CMR is not done in patients with pacemakers) as well as in patients who are claustrophobic [18]. CCTA has the advantage of visualizing coronary anatomy as well as myocardial perfusion and central segments of the pulmonary arteries. Anatomic and functional evaluation of the RV can be performed automatically using cardiac CT (Fig. 12.8) but requires ECG gating, which may increase radiation exposure. Normal values for RV anatomy with CT have been reported [19]. Compared to CMR, CCTA tends to overestimate the end diastolic and end systolic volumes [20]. It is important to recognize that quantitative CMR derived values should not be extrapolated to the CCTA derived measurements [1].

Noncontrast cardiac-CT sequences have the potential of tissue characterization similar to CCTA and can also quantify fatty tissue using difference in the attenuation parameters [21–23].

Advanced postprocessing programs permit quantification of the amount of fat in the entire RV free wall, since CCTA is a 3D methodology,

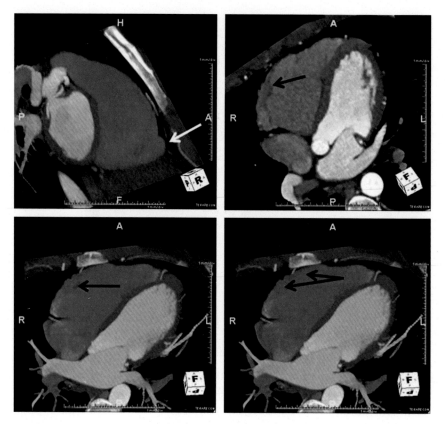

FIGURE 12.8   Cardiac CTA demonstrates RV enlargement and regional RV wall motion abnormalities in a patient with proven ARVC/D.

the entire organ can be analyzed with subsequent development of the parametric maps demonstrating fat distribution in the RV (Fig. 12.9).

Nuclear Imaging Modalities

Multigated Acquisition Scan

Multigated acquisition (MUGA) scan, also known as equilibrium radionuclide angiogram (ERNA) is a noninvasive test used to determine ventricular function. ERNA is a scintigraphic method in which a radiolabeled tracer is intravenously administered to label the patients' red blood cells. Thus it is confined to the blood pool. The duration of the cardiac cycle is determined using ECG gate sensing of the R-Wave and images are acquired when the traces are at equilibrium (Fig. 12.10).

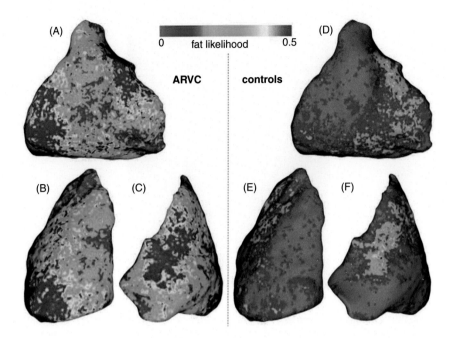

FIGURE 12.9 **Distribution of myocardial fat over the RV free wall in patients with ARVC compared to controls.** Parametric maps are shown in (A, D) lateral, (B, E) inferior, and (C, F) anterior views. *Reproduced with permission from Ref. [24].*

ERNA uses its parametric (synchrony and entropy) data to determine the function of the ventricles. It can successfully differentiate between ARVC/D and a normal heart with a high degree of sensitivity and specificity. It appears to have higher sensitivity than ECHO in detecting ARVC/D but lacks specificity. ERNA is a reproducible technique with minimal interobserver variability. It can be safely performed in patients with an ICD if CMR may be contraindicated [26].

Single Photon Emission Computed Tomography

Single photon emission computed tomography (SPECT) is another nuclear medicine imaging modality. It is able to provide 3D information by offline processing of the images. It is a voxel-based imaging technique. Intravenous injection of radiopharmaceutical (gamma emitting radioisotope) is performed to delineate specific targets in the body to which the injected agents bind.

Mariano-Goulart et al. used gated blood pool SPECT for the diagnosis of ARVC/D [27]. SPECT can help in calculating functional parameters, used for the diagnosis of ARVC/D, but this is not included in the TFC. The latest published evidence to assess the utility of the blood pool SPECT

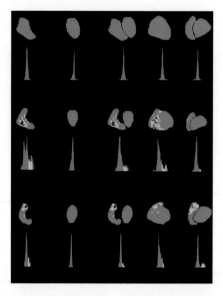

FIGURE 12.10 **Fourier phase images (top) and histograms (bottom) of a patient with normal ERNA (top row), a patient with right ventricular (RV) free wall and subtricuspid area dyskinesia (mid row) and a patient with RV outflow tract dyskinesia (bottom row).** *Reproduced with permission from Ref. [25].*

methodology for the diagnosis of ARVC/D is promising, (Fig. 12.11) [28] but more definitive studies are needed [27].

Metaiodobenzylguanidine

Metaiodobenzylguanidine (MIBG) is a radiolabeled molecule used in the MIBG scan, a scintigraphic method. The iodine radioisotope I-123 or I-131 is used for imaging. It localizes the sympathetic nerve terminals by attaching to norepinephrine. This is an interesting imaging technique which could potentially have a role in diagnosing ARVC/D. Many authors have studied and discussed the role of the sympathetic system activation/involvement in various arrhythmias especially of ventricular origin [29,30]. Different uptake patterns of MIBG are observed in different disease states including myocardial infarction, cardiomyopathies (dilated, hypertrophic), cardiac transplant, and in the long QT syndrome [31].

Wichter et al. demonstrated that tracer (MIBG) uptake is decreased ($p < 0.002$) in the basal septum, posterior and posteroseptal segments in patients with ARVC/D. A cause–effect relationship has not been found but there are numerous hypotheses that either tracer uptake is reduced or tracer uptake is normal or increased but it is cleared from the circulation at a faster rate

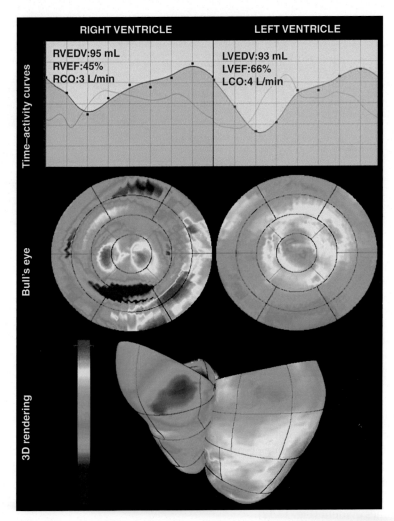

FIGURE 12.11 **Utility of gated blood pool SPECT in clinical routine.** Results obtained with a TOMPOOL software: time-activity curve fit with an 8 points reference curve, image of the regional EF with bull's eye with 17 segments, and 3D-representation at end systole. In this example, the left ventricle function was normal but TOMPOOL revealed an impairment of the right ventricular function, which was confirmed by MRI. *Reproduced with permission from Ref. [28].*

[31]. The cardiac sympathetic nervous system follows the course of the coronary arteries from basal to apical direction in the epicardium or in the subepicardium. During the course of cardiac sympathetic fibers, small branches arise traversing transmurally to innervate the subendocardium [32]. It is well

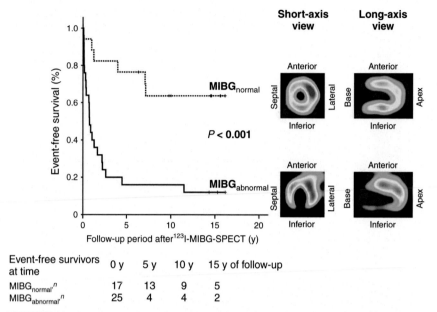

FIGURE 12.12 **Event-free survival during years of follow-up of study population with normal (MIBG normal) versus abnormal (MIBG abnormal) 123I-MIBG uptake.** At right are typical 123I-MIBG images. *Reproduced with permission from Ref. [34].*

known from the pathophysiology of ARVC/D that fibrofatty infiltration of the right ventricle is first present in the epicardium and progresses toward the endocardium [33]. During progression of the disease, sympathetic fibers are damaged which could potentially decrease tracer uptake. Hence, MIBG has the potential to noninvasively diagnose or indicate the risk of sudden cardiac death, since it can detect decreased tracer uptake before the disease becomes evident (Fig. 12.12).

Positron Emission Tomography-Computed Tomography

Positron emission tomography-computed tomography (PET/CT) uses fluorodeoxyglucose (FDG), a glucose analog. Its uptake is increased during inflammation. The clinical and imaging data in patients with suspected ARVC/D are controversial. An imaging modality, at the molecular level could provide important additional information. As an example, cardiac sarcoidosis and ARVC/D have a similar structural and functional presentation and sarcoidosis can mimic ARVC/D. ARVC/D is never associated with florid metabolic uptake in contrast to cardiac sarcoidosis, which may have a high metabolic uptake. FDG PET/CT can help to establish a diagnosis in a challenging situation where inflammation plays a role [35].

Electroanatomic Voltage Mapping (EVM)

Syncope, VT, right ventricular dysfunction, electrophysiological study with inducibility at programmed ventricular stimulation (PVS) are suggested predictors of SCD. EVM accurately characterizes the presence, location, and extent of areas of fibrofatty infiltration (hallmark lesions) in the RV myocardium, which are seen as low-voltage regions called electroanatomical scars (EAS) [36]. Studies have documented that EVM appears to be more sensitive in diagnosing RV scar (fibrofatty infiltration) when compared with CMR [37]. The clinical utility of EVM in detecting scar is limited due to its invasive nature and high cost. Distribution of abnormal mapping also correlates well with severe atrophy and aneurysmal dilatation found at autopsy in hearts in patients with ARVC/D who died suddenly [38].

An abnormal EVM is an independent predictor of an arrhythmic event. Ventricular tachyarrhythmias in ARVC/D patients arise due to right ventricular scarring as a result of fibrofatty infiltration of the myocardium. EVM not only assists in the diagnosis but also in the management of the arrhythmias using voltage mapping guided catheter ablation, to interrupt the reentry circuit of the arrhythmias. Since scar begets arrhythmia, EVM can identify abnormal mapping. It has prognostic value for serious arrhythmia when compared with the traditional indices for RV dilatation/dysfunction. Unipolar EVM is more sensitive than bipolar EVM in detecting RV scar since it is superior in identifying epicardial and midmyocardial scar, which are commonly present in ARVC/D. It can unmask areas of abnormal mapping when bipolar EVM is normal [39].

ADVANCED AND FUSION METHODOLOGIES

CMR is an important tool for functional assessment of the right ventricle. The first step in photoprocessing, is automated or semi-automated RV segmentation on CMR (or CCTA) images. Indistinct cavity borders due to variation in blood flow, wall irregularities (trabeculations), difficulty in discerning the myocardium from nearby structures and complex RV shape make it challenging. Hence, RV segmentation is performed manually, which leads to intra- and interobserver variability [40]. Automated

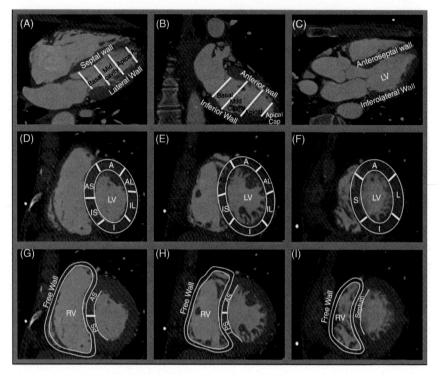

FIGURE 12.13   **Example of the LV and RV segmentation including the 17-segment model.** *Reproduced with permission from Ref. [12].*

algorithms are being developed that provide better results when compared with semiautomated methods (Fig. 12.13). RV endocardium and epicardium segmentation is important for functional assessment and both semiautomated as well as automatic algorithms provide equivalent results. Automated methods do not have any intra- or interobserver variability in comparison with the semiautomated methods (Fig. 12.14) [41].

RV segmentation is also feasible in cardiac CT but the functional and morphological parameters for RV may be different, when compared with echocardiography or CMR[43].

In summary, different imaging modalities have the potential to diagnose ARVC/D but require further research. We anticipate development of completely automated computer programs in the near future that will assess the entire DICOM datasets for quality, perform segmentation as well as global and regional functional analysis.

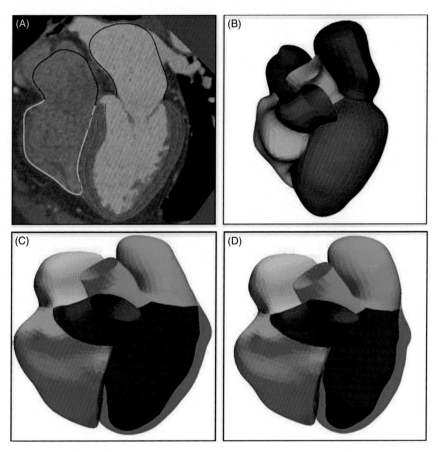

FIGURE 12.14 **Automated heart chamber segmentation.** (A) Four-chamber segmentation with multiplanar reformatting. (B–D) 3D reconstruction of reformatted four-chamber segmentation (B) at end-diastolic phase (C) and end-systolic phase (D). *Reproduced with permission from Ref. [42].*

References

[1] Marcus FI, McKenna WJ, Sherrill D, Basso C, Bauce B, Bluemke DA, Calkins H, Corrado D, Cox MG, Daubert JP, Fontaine G, Gear K, Hauer R, Nava A, Picard MH, Protonotarios N, Saffitz JE, Sanborn DM, Steinberg JS, Tandri H, Thiene G, Towbin JA, Tsatsopoulou A, Wichter T, Zareba W. Diagnosis of arrhythmogenic right ventricular cardiomyopathy/ dysplasia: proposed modification of the task force criteria. Circulation 2010;121:1533–41.

[2] Valsangiacomo Buechel ER, Mertens LL. Imaging the right heart: The use of integrated multimodality imaging. Eur Heart J 2012;33:949–60.

[3] Limongelli G, Rea A, Masarone D, Francalanci MP, Anastasakis A, Calabro R, Maria Giovanna R, Bossone E, Elliott PM, Pacileo G. Right ventricular cardiomyopathies: a multidisciplinary approach to diagnosis. Echocardiography 2015;32(Suppl.1):S75 94.

[4] Prakasa KR, Dalal D, Wang J, Bomma C, Tandri H, Dong J, James C, Tichnell C, Russell SD, Spevak P, Corretti M, Bluemke DA, Calkins H, Abraham TP. Feasibility and variability

of three dimensional echocardiography in arrhythmogenic right ventricular dysplasia/cardiomyopathy. Am J Cardiol 2006;97:703–9.

[5] Sorrell VL, Kumar S, Kalra N. Cardiac imaging in right ventricular cardiomyopathy/dysplasia--how does cardiac imaging assist in understanding the morphologic, functional, and electrical changes of the heart in this disease? J Electrocardiol 2009;42. 137 e1-10.

[6] Prakasa KR, Wang J, Tandri H, Dalal D, Bomma C, Chojnowski R, James C, Tichnell C, Russell S, Judge D, Corretti M, Bluemke D, Calkins H, Abraham TP. Utility of tissue doppler and strain echocardiography in arrhythmogenic right ventricular dysplasia/cardiomyopathy. Am J Cardio 2007;100:507–12.

[7] Korinek J, Wang J, Sengupta PP, Miyazaki C, Kjaergaard J, McMahon E, Abraham TP, Belohlavek M. Two-dimensional strain--a doppler-independent ultrasound method for quantitation of regional deformation: validation *in vitro* and *in vivo*. J Am Soc Echocardiogr 2005;18:1247–53.

[8] Park JH, Kusunose K, Motoki H, Kwon DH, Grimm RA, Griffin BP, Marwick TH, Popovic ZB. Assessment of right ventricular longitudinal strain in patients with ischemic cardiomyopathy: Head-to-head comparison between two-dimensional speckle-based strain and velocity vector imaging using volumetric assessment by cardiac magnetic resonance as a "gold standard". Echocardiography 2015;32:956–65.

[9] Auger DA, Zhong X, Epstein FH, Spottiswoode BS. Mapping right ventricular myocardial mechanics using 3d cine dense cardiovascular magnetic resonance. J Cardiovasc Magn Reson 2012;14:4.

[10] Kang Y, Cheng L, Li L, Chen H, Sun M, Wei Z, Pan C, Shu X. Early detection of anthracycline-induced cardiotoxicity using two-dimensional speckle tracking echocardiography. Cardiol J 2013;20:592–9.

[11] Maron BJ, Chaitman BR, Ackerman MJ, Bayes de Luna A, Corrado D, Crosson JE, Deal BJ, Driscoll DJ, Estes NA III, Araujo CG, Liang DH, Mitten MJ, Myerburg RJ, Pelliccia A, Thompson PD, Towbin JA, Van Camp SP. Recommendations for physical activity and recreational sports participation for young patients with genetic cardiovascular diseases. Circulation 2004;109:2807–16.

[12] Sato PY, Musa H, Coombs W, Guerrero Serna G, Patino GA, Taffet SM, Isom LL, Delmar M. Loss of plakophilin-2 expression leads to decreased sodium current and slower conduction velocity in cultured cardiac myocytes. Circ Res 2009;105:523–6.

[13] van Tintelen JP, Entius MM, Bhuiyan ZA, Jongbloed R, Wiesfeld AC, Wilde AA, van der Smagt J, Boven LG, Mannens MM, van Langen IM, Hofstra RM, Otterspoor LC, Doevendans PA, Rodriguez LM, van Gelder IC, Hauer RN. Plakophilin-2 mutations are the major determinant of familial arrhythmogenic right ventricular dysplasia/cardiomyopathy. Circulation 2006;113:1650–8.

[14] Thiene G, Nava A, Corrado D, Rossi L, Pennelli N. Right ventricular cardiomyopathy and sudden death in young people. New Engl J Med 1988;318:129–33.

[15] Perrin MJ, Angaran P, Laksman Z, Zhang H, Porepa LF, Rutberg J, James C, Krahn AD, Judge DP, Calkins H, Gollob MH. Exercise testing in asymptomatic gene carriers exposes a latent electrical substrate of arrhythmogenic right ventricular cardiomyopathy. J Am Coll Cardiol 2013;62:1772–9.

[16] La Gerche A, Claessen G, Dymarkowski S, Voigt JU, De Buck F, Vanhees L, Droogne W, Van Cleemput J, Claus P, Heidbuchel H. Exercise-induced right ventricular dysfunction is associated with ventricular arrhythmias in endurance athletes. Eur Heart J 2015.

[17] Bomma C, Dalal D, Tandri H, Prakasa K, Nasir K, Roguin A, Piccini J, Dong J, Mahadevappa M, Tichnell C, James C, Lima JA, Fishman E, Calkins H, Bluemke DA. Evolving role of multidetector computed tomography in evaluation of arrhythmogenic right ventricular dysplasia/cardiomyopathy. Am J Cardiol 2007;100:99–105.

[18] Romero J, Mejia-Lopez E, Manrique C, Lucariello R. Arrhythmogenic right ventricular cardiomyopathy (arvc/d): a systematic literature review. Clin Med Insights Cardiol 2013;7:97–114.

[19] Lin FY, Devereux RB, Roman MJ, Meng J, Jow VM, Jacobs A, Weinsaft JW, Shaw LJ, Berman DS, Callister TQ, Min JK. Cardiac chamber volumes, function, and mass as determined by 64-multidetector row computed tomography: mean values among healthy adults free of hypertension and obesity. JACC Cardiovasc Imaging 2008;1:782–6.

[20] Plumhans C, Muhlenbruch G, Rapaee A, Sim KH, Seyfarth T, Gunther RW, Mahnken AH. Assessment of global right ventricular function on 64-mdct compared with MRI. AJR Am J Roentgenol 2008;190:1358–61.

[21] Raney AR, Saremi F, Kenchaiah S, Gurudevan SV, Narula J, Narula N, Channual S. Multidetector computed tomography shows intramyocardial fat deposition. J Cardiovasc Comput Tomogr 2008;2:152–63.

[22] Borkan GA, Gerzof SG, Robbins AH, Hults DE, Silbert CK, Silbert JE. Assessment of abdominal fat content by computed tomography. Am J Clin Nutr 1982;36:172–7.

[23] Tada H, Shimizu W, Ohe T, Hamada S, Kurita T, Aihara N, Kamakura S, Takamiya M, Shimomura K. Usefulness of electron-beam computed tomography in arrhythmogenic right ventricular dysplasia. Circulation 1996;94:437–44.

[24] Cochet H, Denis A, Komatsu Y, Jadidi AS, Ait Ali T, Sacher F, Derval N, Relan J, Sermesant M, Corneloup O, Hocini M, Haissaguerre M, Laurent F, Montaudon M, Jais P. Automated quantification of right ventricular fat at contrast-enhanced cardiac multidetector ct in arrhythmogenic right ventricular cardiomyopathy. Radiology 2015;275(3):683–91.

[25] Rouzet F, Sarda-Mantel L, Lebtahi R, Dinanian S, Frank R, Daou D, Leenhardt A, Slama MS, Le Guludec D. Early detection of right ventricular functional abnormalities in patients with complex right premature ventricular contractions. Nucl Med Commun 2008;29:901–6.

[26] Johnson CJ, Roberts JD, James JH, Hoffmayer KS, Badhwar N, Ku IA, Zhao S, Naeger DM, Rao RK, O'Connell JW, De Marco T, Botvinick EH, Scheinman MM. Comparison of radionuclide angiographic synchrony analysis to echocardiography and magnetic resonance imaging for the diagnosis of arrhythmogenic right ventricular cardiomyopathy. Heart Rhythm 2015;12:1268–75.

[27] Mariano-Goulart D, Dechaux L, Rouzet F, Barbotte E, Caderas de Kerleau C, Rossi M, Le Guludec D. Diagnosis of diffuse and localized arrhythmogenic right ventricular dysplasia by gated blood-pool spect. J Nucl Med 2007;48:1416–23.

[28] Dercle L, Giraudmaillet T, Pascal P, Lairez O, Chisin R, Marachet MA, Ouali M, Rousseau H, Bastie D, Berry I. Is tompool (gated blood-pool spect processing software) accurate to diagnose right and left ventricular dysfunction in a clinical setting? J Nucl Cardiol 2014;21:1011–22.

[29] Lemery R, Brugada P, Janssen J, Cheriex E, Dugernier T, Wellens HJ. Nonischemic sustained ventricular tachycardia: clinical outcome in 12 patients with arrhythmogenic right ventricular dysplasia. J Am Coll Cardiol 1989;14:96–105.

[30] Leclercq JF, Coumel P. Characteristics, prognosis and treatment of the ventricular arrhythmias of right ventricular dysplasia. Eur Heart J 1989;10(Suppl. D):61–7.

[31] Wichter T, Hindricks G, Lerch H, Bartenstein P, Borggrefe M, Schober O, Breithardt G. Regional myocardial sympathetic dysinnervation in arrhythmogenic right ventricular cardiomyopathy. An analysis using 123i-meta-iodobenzylguanidine scintigraphy. Circulation 1994;89:667–83.

[32] Sisson JC, Lynch JJ, Johnson J, Jaques S Jr, Wu D, Bolgos G, Lucchesi BR, Wieland DM. Scintigraphic detection of regional disruption of adrenergic neurons in the heart. Am Heart J 1988;116:67–76.

[33] Strain J. Adipose dysplasia of the right ventricle: Is endomyocardial biopsy useful? Eur Heart J 1989;10(Suppl. D):84–8.

[34] Paul M, Wichter T, Kies P, Gerss J, Wollmann C, Rahbar K, Eckardt L, Breithardt G, Schober O, Schulze-Bahr E, Schafers M. Cardiac sympathetic dysfunction in genotyped patients with arrhythmogenic right ventricular cardiomyopathy and risk of recurrent ventricular tachyarrhythmias. J Nucl Med 2011;52:1559–65.

[35] Choo WK, Denison AR, Miller DR, Dempsey OJ, Dawson DK, Broadhurst PA. Cardiac sarcoid or arrhythmogenic right ventricular cardiomyopathy: A role for positron emission tomography (pet)? J Nucl Cardiol 2013;20:479–80.

[36] Silvano M, Corrado D, Kobe J, Monnig G, Basso C, Thiene G, Eckardt L. Risk stratification in arrhythmogenic right ventricular cardiomyopathy. Herzschrittmachertherapie & Elektrophysiologie 2013;24:202–8.

[37] Marra MP, Leoni L, Bauce B, Corbetti F, Zorzi A, Migliore F, Silvano M, Rigato I, Tona F, Tarantini G, Cacciavillani L, Basso C, Buja G, Thiene G, Iliceto S, Corrado D. Imaging study of ventricular scar in arrhythmogenic right ventricular cardiomyopathy: Comparison of 3d standard electroanatomical voltage mapping and contrast-enhanced cardiac magnetic resonance. Circ Arrhythm Electrophysiol 2012;5:91–100.

[38] Basso C, Thiene G, Corrado D, Angelini A, Nava A, Valente M. Arrhythmogenic right ventricular cardiomyopathy. Dysplasia, dystrophy, or myocarditis? Circulation 1996;94:983–91.

[39] Migliore F, Zorzi A, Silvano M, Bevilacqua M, Leoni L, Marra MP, Elmaghawry M, Brugnaro L, Dal Lin C, Bauce B, Rigato I, Tarantini G, Basso C, Buja G, Thiene G, Iliceto S, Corrado D. Prognostic value of endocardial voltage mapping in patients with arrhythmogenic right ventricular cardiomyopathy/dysplasia. Circ Arrhythm Electrophysiol 2013;6:167–76.

[40] Caudron J, Fares J, Lefebvre V, Vivier PH, Petitjean C, Dacher JN. Cardiac MRI assessment of right ventricular function in acquired heart disease: factors of variability. Acad Radio 2012;19:991–1002.

[41] Petitjean C, Zuluaga MA, Bai W, Dacher JN, Grosgeorge D, Caudron J, Ruan S, Ayed IB, Cardoso MJ, Chen HC, Jimenez-Carretero D, Ledesma-Carbayo MJ, Davatzikos C, Doshi J, Erus G, Maier OM, Nambakhsh CM, Ou Y, Ourselin S, Peng CW, Peters NS, Peters TM, Rajchl M, Rueckert D, Santos A, Shi W, Wang CW, Wang H, Yuan J. Right ventricle segmentation from cardiac MRI: a collation study. Med Image Anal 2015;19:187–202.

[42] Rizvi A, Deano RC, Bachman DP, Xiong G, Min JK, Truong QA. Analysis of ventricular function by ct. J Cardiovasc Comput Tomogr 2015;9:1–12.

[43] Coche F, Walker MJ, Zech F, de Crombrugghe R, Vlassenbroek A. Quantitative right and left ventricular functional analysis during gated whole-chest mdct: A feasibility study comparing automatic segmentation to semi-manual contouring. Eur J Radiol 2010;74:e138–43.

13

Selected Clinical Cases from Our Practice

Aiden Abidov, Ahmed K. Pasha, Arun Kannan,
Isabel B. Oliva, Frank I. Marcus

Department of Medicine/Division of Cardiology
and Department of Medical Imaging,
University of Arizona, Tucson, AZ, USA

CASE 1

A 22-year-old man noticed palpitations while hiking. He became lightheaded and then lost consciousness for several minutes. Emergency Medical Services (EMS) arrived at the scene. His ECG showed a wide complex tachycardia at 180 beats/min with left bundle branch block (LBBB) morphology.

He was cardioverted to normal sinus rhythm. A 12-lead ECG showed inverted T waves in leads V1, V2, and V3. He was brought to the hospital for evaluation. He denied a previous history of syncope. Past medical history and social history were noncontributory. There was no known familial cardiomyopathy.

Cardiac CTA showed that his coronary arteries were normal. However, CTA showed right ventricular (RV) enlargement and focal RV wall motion abnormalities. CMR confirmed these findings and demonstrated marked RV dilatation with akinesis, dyskinesis, and aneurysmal dilation of the RV free wall (Fig. 13.1). The RV end-diastolic volume index was 167 mL/m^2 (the normal value by the Task Force criteria (TFC) is > 100 mL/m^2) and RV ejection fraction (RVEF) was 36% (normal TFC is > 45%).

Outcomes

The patient met TFC for ARVC/D. The patient was treated with an implanted cardioverter defibrillator (ICD).

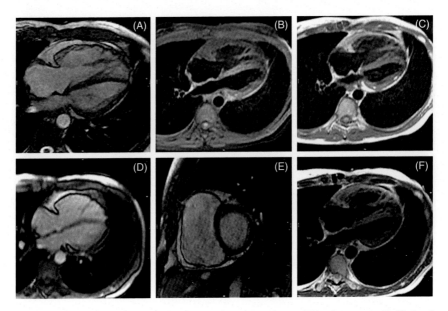

FIGURE 13.1   **CMR images of the patient in Case 1.** Severe RV enlargement (A–F), focal RV free wall aneurysms (A, D) and evidence of the RV and LV fibrofatty changes (B, C) is noted.

Teaching point: This is a typical ARVC/D case presenting with syncope due to rapid VT. The patient was successfully diagnosed by CTA and underwent subsequent ICD implantation.

CASE 2

A 72-year-old woman whose children were diagnosed with ARVC/D presented for evaluation.

She had no significant complaints. She was physically active, and had no history of syncope, palpitations, dyspnea, or chest pain. Her son had been resuscitated from sudden death during exercise at the age of 51. He was diagnosed with ARVC/D and was found to have a genetic mutation in the plakophilin gene.

Another son was also found to be genotype-positive for ARVC/D.

The woman was found to be a mutation in PKP2 c.2146-1G>C (heterozygous) for ARVC/D.

Her physical examination was unremarkable. An ECG showed sinus rhythm without any ST/T wave abnormalities. A transthoracic echocardiogram showed a normal left ventricular (LV) size, the LV ejection fraction (LVEF) was 52%, the RV size was normal with mildly reduced systolic function.

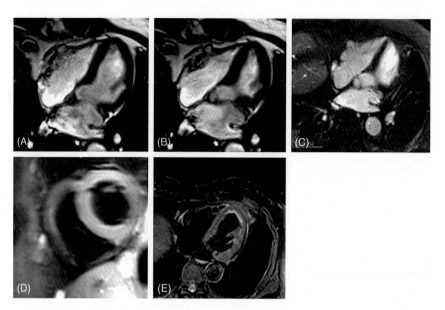

FIGURE 13.2 **CMR images of the patient in Case 2.** RV enlargement, RV free wall aneurysm (A–C). Triple-inversion recovery images demonstrate fibro-fatty changes in the thinned RV inferior wall (D) and subtricuspid area (E). Fibro-fatty changes were also noted in the left ventricle (septum and lateral segment) (E).

Her CMR was performed at an outside hospital and was reported to be normal.

We reviewed this CMR (Fig. 13.2) and it showed RV enlargement, depressed systolic RV function with an RVEF of 45%, and regional motion abnormalities in the RV free wall including a focal aneurysm in the mid-RV free wall and akinesia in the subtricuspid area. Additionally, the RV free wall was thin and there were fibrofatty changes both of the RV and the LV in the septum (septal and lateral wall).

Analysis

The patient fulfilled TFC for ARVC/D and had CMR evidence of RV and LV dysplasia that was not reported by the previous interpretation of the images.

An ICD was implanted and the patient was advised to follow-up with her cardiologist and electrophysiologist.

Teaching point: Patients with personal or family history of SCD and established diagnosis of ARVC/D should have a CMR examination for evaluation of RV and LV involvement. CMR for the patient with known or suspected ARVC/D should be performed using standardized protocols and reviewed by a specialist with expertise in the field of ARVC/D. Implantation of an ICD in this elderly asymptomatic woman who had normal ECG and no electrical abnormalities is controversial.

CASE 3

A 35-year-old man presented to the cardiology clinic for evaluation of ARVC/D. He had exercise-induced asthma, obstructive sleep apnea treated with continuous positive airway pressure (CPAP), lateral epicondylitis, allergies, and gastroesophageal reflux (GERD).

The patient had a family history of sudden death. His maternal uncle died suddenly at the age of 53. The uncle's son was found to have an ARVC/D-associated gene while undergoing evaluation for recurrent syncope/collapse. Therefore this patient requested evaluation.

The patient denied syncope, but had occasional exercise-induced dizziness and palpitations. His initial resting blood pressure was 157/90, but on subsequent visits this was found to be normal without any antihypertensive medications. His physical examination was unremarkable.

His ECG showed sinus rhythm. The T waves were upright in the precordial leads, and there was no evidence of arrhythmia. He had a 30-day event monitor and there were no premature ventricular contractions.

A two-dimensional transthoracic echocardiogram (TTE) was performed that showed a normal RV size. There was no echocardiographic evidence for RV hypertrophy. The tricuspid annular plane of systolic excursion (TAPSE) was measured as 22 mm (normal value >16 mm). The visually estimated RV global systolic function was normal. No definite pathological RV regional wall motion abnormalities were seen.

CMR showed no evidence of focal RV wall motion abnormalities and a normal RV size and systolic function (Fig. 13.3). Prominent trabeculations were seen in the RV (possibly a normal variant). There was no trabecular disarray. The RV end-diastolic volume was 88 mL/m^2. RV cardiac output was normal at 5.4 L/min. The calculated RVEF was 50%. The LV was normal structurally and functionally.

Evlauations

Since there was no clinical or phenotypic evidence of ARVC/D based on ECG, CMR, or echocardiogram, the patient was advised to have a yearly follow-up with his cardiologist with sequential Holter monitoring and echocardiograms.

Teaching point: This patient had a family member who was gene positive. His imaging data did not support the diagnosis of ARVC/D. A few CMR values may be considered of borderline clinical significance (prominent RV trabeculation; borderline RV size). Serial Holter and imaging studies are recommended every 2–3 years if asymptomatic. The reason for adding Holter studies is that in many patients,

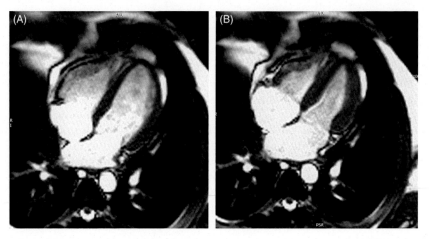

FIGURE 13.3   **CMR images of the patient in Case 3.** Diastolic (A) and systolic (B) frames are shown, demonstrating normal global and regional RV function.

ECG manifestations of the disease precede structural abnormalities. The patient does not perform competitive athletic activities but excessive/strenuous exercise should be avoided.

CASE 4

A 48-year-old man was seen in the electrophysiology (EP) clinic. He had intermittent chest pain and had LBBB on his ECG, prompting his primary care physician (PCP) to refer him for further cardiac evaluation. Coronary angiography revealed nonobstructive coronaries. He had a history of hiatal hernia.

The patient was physically active and walked for about an hour a day. His father died at the age of 42 due to a heart attack.

A TTE showed normal LV and RV function. His Holter monitor showed sinus rhythm with frequent episodes of an accelerated idioventricular rhythm.

His CMR showed a dilated RV with mildly decreased systolic function (RVEF = 39.5%). There was focal mid-to-distal RV free wall akinesis as well as focal dyskinetic changes in the RV apex. There were a few fibrotic areas in the basal to midinferoseptal wall of the LV as well as in the RV midfree wall and RV apex. Significant RV trabeculation and trabecular disarray were also noted.

No abnormal desmosomal genes compatible with ARVC/D were identified.

Evaluation

The patient had no history of syncope. There was a history of SCD in his family but the diagnostic cause was unknown. An ICD was not indicated. He was treated with sotalol and advised to follow-up with his cardiologist with ECG, Holter, and serial CMR every 2–3 years.

He did not have TFC for ARVC/D.

Teaching point: The patient had an abnormal ventricular rhythm but no clinical findings or history to suggest ARVC/D. The ECG of LBBB is unusual in a patient with ARVC/D. His genetic testing was negative but CMR was borderline for ARVC/D (only one TFC met). Normal genetic testing does not exclude the presence of ARVC/D. There is a high index of suspicion of ARVC/D in this patient. Holter monitoring and imaging is suggested (please see comment in Case #3) with close follow-up by cardiology. The patient should avoid competitive sports.

CASE 5

A 75-year-old man with known ARVC/D and recurrent VT presented to the EP clinic for a follow-up. The patient denied episodes of syncope in his family. His uncle died with an unknown rhythm at the age of 70. The patient was physically active and did not have any significant health problems. A 12-lead ECG showed sinus rhythm with T wave inversion from leads V1 through V4.

A TTE (Fig. 13.4) showed LVEF of 40–49% (normal value ≥55%), severely dilated RV, moderately reduced RV systolic function, and a severely dilated right atrium. He had increased RV trabeculation and RV trabecular disarray.

Evaluation

The patient had a history of recurrent VT and was diagnosed with ARVC/D many years ago. He refused ICD implantation. Despite significant imaging findings, and especially LV dysfunction, the patient maintains a good quality of life, exercises regularly (jogging), and did not feel the need to undergo further treatment.

Teaching point: Not all patients with ARVC/D require ICD placement. This patient would have been eligible for ICD but he refused. Mortality among patients with ARVC/D is highest among young patients, with considerable decrease in mortality among elderly ARVC/D patients. We cannot accurately predict individualized mortality but high-risk imaging features suggestive of adverse outcomes are severe

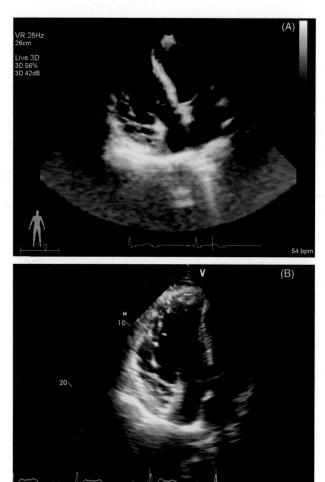

FIGURE 13.4 **Echo images of the patient in Case 5.** Dilated RV and increased RV trabeculations and trabecular disarray in the proximal (A) and distal (B) RV.

RV dilatation and failure; significant LV involvement; and a history of ventricular tachycardia. Patients should be encouraged to avoid strenuous exercise.

CASE 6

A 48-year-old woman with ARVC/D diagnosed 10 years ago was seen in the EP clinic for follow-up. She denied a familial history of syncope or SCD.

A 12-lead ECG showed inverted T waves in leads V1–V5.

CMR showed marked RV dilatation and akinesis of a large portion of the RV free wall, RV outflow tract, and basal inferior RV. There were also areas of fibrosis in the lateral wall of the LV.

A Holter monitor study done a few years prior showed one short burst of nonsustained ventricular tachycardia. A repeat Holter monitor revealed multiple PVCs as well as a run of nonsustained ventricular tachycardia.

She was advised to follow-up with cardiology for ICD placement. However, the patient refused the device. Beta-blockers were recommended, but the patient was reluctant to accept this treatment due to concern of adverse effects.

Evaluation

This asymptomatic woman with ARVC/D refused an ICD. She was advised to have regular follow up with an electrophysiologist to determine possible progression of the frequency of PVCs or symptomatic VT.

Teaching point: The patient has a known history of ARVC/D. She remains asymptomatic and refuses ICD; however, careful review of her CMR revealed LV involvement and this may have an adverse prognosis. Treatment with beta-blockers may be suggested for patients with borderline or mild symptoms but its efficacy is unknown.

CASE 7

A 45-year-old man was evaluated at a hospital after an episode of syncope while on a bicycle ride. A 12-lead ECG revealed T wave inversion in V1–V5, and frequent PVCs with LBBB morphology and superior QRS axis. Due to inverted T waves in multiple precordial leads, there was concern for ARVC/D and a CMR was performed at an outside hospital.

The CMR was reported to show normal RV size and global function. Subsequent review of the images showed hypokinesis at the RV apex without evidence of fibrofatty infiltration. He was advised to take beta-blockers but refused.

Two years later, the patient had another episode of syncope, while bicycling had cardiac arrest, and died. An autopsy revealed that the RV was diffusely replaced by fibrofatty tissue as well as with subepicardial fat in the lateral and posterior walls of the LV (Fig. 13.5).

Evaluation

Imaging criteria by TFC did not support the diagnosis of ARVC/D. However, the pathology at autopsy was diagnostic for the presence of this disease.

FIGURE 13.5   **Pathology images of the patient in Case 7.** Significant biventricular fibro-fatty changes consistent with ARVC/D.

Teaching point: ARVC/D is a diagnosis with significant associated rate of fatality and this should be clearly discussed with the patients. Even though the ECG of this patient was definitely abnormal (T wave inversion V1-V5), the patient's physician relied on the interpretation of the CMR rather than having the imaging study interpreted by the physician experienced with this disease.

Subject Index

Printed in the United States
By Bookmasters